11550

27.95/18.17
mcs

D0606324

629.41 MAL

11550

Mallove, Eugene F.
The starflight handbook
LACC S

$27.95

LACOMBE COMPOSITE
HIGH SCHOOL LIBRARY

THE
STARFLIGHT HANDBOOK

LACOMBE COMPOSITE
HIGH SCHOOL LIBRARY

The Wiley Science Editions

THE STARFLIGHT HANDBOOK

A Pioneer's Guide to Interstellar Travel

Eugene F. Mallove
and
Gregory L. Matloff

WILEY SCIENCE EDITIONS

John Wiley & Sons, Inc.
New York • Chichester • Brisbane • Toronto • Singapore

Publisher: Stephen Kippur
Editor: David Sobel
Managing Editor: Frank Grazioli
Editing, Design, & Production: G&H SOHO, Ltd.

This publication is designed to provide accurate and authoritative information in regard to the subject matter covered. It is sold with the understanding that the publisher is not engaged in rendering legal, accounting, or other professional service. If legal advice or other expert assistance is required, the services of a competent professional person should be sought. FROM A DECLARATION OF PRINCIPLES JOINTLY ADOPTED BY A COMMITTEE OF THE AMERICAN BAR ASSOCIATION AND A COMMITTEE OF PUBLISHERS.

Copyright © 1989 by Eugene F. Mallove and Gregory L. Matloff

All rights reserved. Published simultaneously in Canada.

Reproduction or translation of any part of this work beyond that permitted by section 107 or 108 of the 1976 United States Copyright Act without the permission of the copyright owner is unlawful. Requests for permission or further information should be addressed to the Permission Department, John Wiley & Sons, Inc.

Library of Congress Cataloging-in-Publication Data

Mallove, Eugene F.
　　The starflight handbook : a pioneer's guide to interstellar travel
　　Eugene F. Mallove and Gregory L. Matloff.
　　　　p.　　c.m. — (Wiley science editions)
　　Bibliography: p.
　　Includes index.
　　ISBN 0-471-61912-4
　　1. Interstellar travel.　　I. Matloff, Gregory L.　　II. Title.
III. Series.
TL790.M26　　　1989
629.4′1—dc19　　　　　　　　　　　　　　　　　　　　　　　　　　　88-31933
　　　　　　　　　　　　　　　　　　　　　　　　　　　　　　　　　　　CIP

Printed in the United States of America

89　90　10　9　8　7　6　5　4　3　2　1

Dedicated to the memory of
Robert H. Goddard
and
Konstantin E. Tsiolkovskii

and to all
who share and who will share
their magical dreams.

Too low they build, who build beneath the stars

Edward Young, 1683–1765, *Night Thoughts*
(Inscribed on a wall of the Library of Congress)

Oh, write of me not "Died in bitter pains,"
But "emigrated to another star!"

Helen Hunt Jackson, 1831–1885, *Emigravit*

Preface

The first interstellar spacecraft, Pioneer 10, departed Earth in March 1972. As we complete this book in 1988, it has been five years since Pioneer 10 overstepped the boundaries of the Solar System on an endless journey among the stars. What better time to complete a project conceived nearly a decade ago—the publication in book form of the planet's first compilation of scientific and engineering knowledge about interstellar travel.

The vast distances that separate the stars loom as a barrier even to automated reconnaissance missions to nearby planetary systems. One-way trips to the nearer stars shorter than hundreds or even thousands of years are feasible only with very advanced technology now on the horizon. Even so, precursor missions to explore near interstellar space will probably happen by the mid-twenty first century, if not sooner. Much faster probes and later missions bearing people will be launched toward the stars when advanced propulsion systems now only theoretically possible come to fruition.

This book is for those who are convinced or who need to be convinced that bridging the interstellar gulf is an attainable goal and represents perhaps the greatest technical challenge for humankind. But how to extend our reach over a chasm that at its most narrow is 270,000 times the distance from the Sun to the Earth? Necessarily, the bulk of the text addresses the technology of propulsion systems—both current and proposed—that we might apply to starflight. Not overlooked, however, are discussions of interstellar navigation, relativistic effects in starflight, effects of the interstellar medium, scientific payloads, interstellar arks, intelligent computers to guide the missions, and interstellar communication systems for returning data.

We did not intend this to be a "textbook" on interstellar flight—the field is far too vast and technically diverse for a book of this size to boast that. We believe instead that this will serve as a good summary compendium of organized interstellar concepts, formulae, and referenced material. Readers should think of the book as a challenging platform from which to leap into the vast literature of starflight—a working guide for the would-be star traveler.

If you are not fluent in mathematics, fear not. Wherever possible, we have separated mathematical and other detailed technical developments from the main text, relegating them to accessible Technical Notes that can be glanced over. Other than these hopefully inviting elaborations, we presuppose no more than a rudimentary acquaintance with mathematics. However, a willingness to deal with "powers of ten"—that is, scientific notation—is a must. Hence the tutorial in Appendix 1, if you need it.

We are enormously indebted to the multitude of scientists, engineers, and dreamers, living and dead, named and nameless, without whose vision and efforts this book would have been impossible. We are particularly grateful for fellowship with our colleagues in the British Interplanetary Society who have nurtured the dream of starflight for so many years, among them, notably, Robert L. Forward, Anthony Martin, and Alan Bond. Nor would this book have emerged without the direction of editor David Sobel at John Wiley & Sons, the able assistance of Nancy Woodruff, Nana Clark, Frank Grazioli, and David Sassian, and the encouragement of our literary agent, Richard Curtis. Special thanks are due to artist Constance Bangs, who created the chapter frontispieces. Her work, which usually celebrates the Earth, extends here to the realm of the extraterrestrial. A note of gratitude as well to Mother Nature for arranging Star Island off the coast of New Hampshire and Salt Island in Long Island Sound, two inspirational gems that helped us in the home stretch.

We hope that this work will be technically useful and an inspiration to this and future generations of starship pioneers who aim at the stars. Our fondest dream is that in the year 2001, the dog-eared first edition will be quaintly obsolete.

Eugene F. Mallove
Gregory L. Matloff

Contents

4 Nuclear Pulse Propulsion

57

5 Beamed Energy Propulsion

71

6 Solar Sail Starships: Clipper Ships of the Galaxy

89

7 Fusion Ramjets

107

1 Introduction to Starflight

There can be no thought of finishing, for "aiming at the stars," both literally and figuratively, is a problem to occupy generations, so that no matter how much progress one makes, there is always the thrill of just beginning . . .

Robert H. Goddard, 1932, in a letter to H. G. Wells

The finer part of mankind will, in all likelihood, never perish—they will migrate from sun to sun as they go out. And so there is no end to life, to intellect and the perfection of humanity. Its progress is everlasting.

Konstantin E. Tsiolkovskii, 1857–1935

Whys and Wherefores

Interstellar travel is real, despite what the doubters say. We'll begin by differing politely but emphatically with that distinguished radio astronomy pioneer from Harvard University, Edward Purcell, who at the dawn of the space age made many classic pessimistic assumptions about how starflight might be accomplished. Then, in 1960, he penned the boldest denial of interstellar flight on record, a distinction of dubious honor:

All this stuff about traveling around the universe in space suits—except for *local* exploration which I have not discussed—belongs back where it came from, on the cereal box.

It was not the first time that an otherwise perceptive scientist had displayed a peculiar failure of imagination about space travel. Witness the remarkable pronouncement by a British scientist in the 1920s about a venture requiring much less expansive thinking:

This foolish idea of shooting at the moon is an example of the absurd length to which vicious specialization will carry scientists. To escape the Earth's gravitation a projectile needs a velocity of 7 miles per second. The thermal energy at this speed is 15,180 calories [per gram]. Hence the proposition appears to be basically impossible.

A.W. Bickerton, 1926

Not only are the first emissaries to the stars already under way (Pioneers 10 and 11 and Voyagers 1 and 2, therefore starflight of an extremely primitive kind is being done right now), but also many thinkers have devoted considerable attention to finding ingenious ways to make trips to the stars by craft much fleeter than these early "slow boats." The plans these visionaries have developed in the past three decades are impressive.

Some plans are extraordinarily far-reaching like the British Interplanetary Society's Daedalus study that envisioned an admittedly "proof-of-concept" thermonuclear pulse rocket that could reach nearby Barnard's star with scientific instruments in about half a century (2). Others, such as the Jet Propulsion Laboratory's study of an interstellar precursor mission, have much more limited objectives: exploring the interstellar medium to a range less than 1% of the distance to the nearest star (3). Yet such a mission could be mounted within 20 years assuming only conservative extensions of technology.

But to be frank: Interstellar travel is not easy. It cannot be accomplished simply by wishing for a convenient wormhole in space-time to drop through to the other side of the universe or for a short hop to a nearby sun. Interstellar travel, "starflight" for short, may never be done by Earthlings in the ways outlined here, although we believe that some or all of these methods will eventually be used. But the inexorable and difficult buildup of technology and science on the platforms of past labors insures a significant place for starflight in humankind's future. Some or all of the propulsion systems described in this book will play a role in taking at first machines (robotic probes), and then people, to the stars.

The big problem with starflight is, of course, distance, and that is why the bulk of this handbook is devoted to methods of interstellar propulsion. The ancillary problems of guidance and navigation, payload content, reliability, and so on, though difficult, are relatively minor issues compared with the primary hurdle of attaining speed sufficient to reduce to tolerable lengths transit times to the stars.

Why would we want to go to the stars in the first place? Is it not enough to be near the one—the nearest star—that lights up our days? Surely, we can study the Sun, study the Solar System's planets, moons, asteroids, and comets. We have already made dramatic progress through millennia of astronomical observations, and much more recently, inter-planetary spaceflight. But soon in cosmic history we will have examined the tiny piece of galactic real estate that is the Solar System and it will be time to move on. We will probably want to begin starflight in earnest even before the Solar System has been completely explored—*just because the stars are there* and since by nature we are driven to be explorers. Or, if you prefer Alfred North Whitehead's dictum, because, "without adventure civilization is in full decay."

What draws us to the stars are not those fusion fires themselves but the planets that attend many if not most suns—at least 50% by some recent informed estimates. As children of a blue-green oasis planet in a wheeling system of at least nine major worlds and a multitude of moons, we yearn to explore those imagined realms so far, far away. Solar System-based instruments are gradually revealing more and more about the probability of the existence of planets surrounding the nearer stars, and it will not be long before we have definite proof of their presence. There is the possibility that we may eventually be able to form crude images from afar of even small Earth-size planets within other solar systems, using highly specialized and very expensive optical techniques (see Chapter 16).

But to sift the sands of these truly remote worlds, to explore those planets and moons for the first glimmerings of life or the remains of extinct life, to sense the beauty of their environments through the dispatch of robot instruments, or from the reports returned by human crews—these are our dreams and the legitimate goals of interstellar travel.

First we must confront a controversy—a genuine and unfortunate though understandable split in the ranks of science. There are many, among them Edward Purcell, who suggest that exploration or colonization by humans of extrasolar planets—or even probing by remote instruments—will never be done because there is a way to accomplish this for "free," albeit vicariously. We can simply cock our radio telescope ears to the heavens, search patiently for signals from other civilizations, and tune into a galactic network of interstellar information. The sensible and noble idea of "conventional" SETI—the search for extraterrestrial intelligence—has great promise and deserves fervent support.

But it is by no means certain that such interstellar signaling will be a prevailing mode of galactic discourse, even though we would like to believe that it is. One can imagine, for example, a civilization of dolphinlike ocean creatures among whom are philosophers, mathematicians, poets, and musicians, but who have not had the means or interest to develop technology. Arthur C. Clarke's words about starflight in *The*

Promise of Space (1967) ring true, "This proxy [robot probe] exploration of the universe is certainly one way in which it would be possible to gain knowledge of star systems which lacked garrulous, radio-equipped inhabitants; it might be the only way."

On the other hand, compelling scientific reasons may arise within the next fifty years to encourage serious thought at least about robotic probes to nearby solar systems. Advanced techniques in optical astronomy (see Chapter 16) may make possible not only the detection of Earth-size planets orbiting nearby stars but also the determination by spectroscopy of the hallmarks of living extrasolar worlds: chemical constituents in planetary atmospheres such as oxygen coexisting with methane. The scientific interest in such worlds would be enormous and would warrant intense scrutiny by SETI researchers. If the examination of the radio spectrum of these nearby "Earthlike worlds" (ELWs) revealed no evidence of technological civilizations, starflight would be the only way to investigate them in detail.

But starflight is much more than a hedge against failure of the many active SETI efforts now ongoing and soon to be inaugurated around the world. Starflight is indeed the ultimate means by which terrestrial life and human culture—of the enlightened sort, it is hoped—can stabilize itself against local astrophysical or even provincial biological catastrophes. By providently spreading terrestrial seeds far beyond this Solar System, we will be insuring the longevity of what has begun on this tiny world.

The interstellar imperative—the "bottom line" of starflight—is that ours should become a civilization that can outlive its star. The life-giving Sun will not always remain benign. About five billion years from now, its core of hydrogen fusion fuel nearly spent, the Sun will expand in an angry "red giant" phase, incinerating all the creations of humanity that might still exist in the inner Solar System. There are other possible astrophysical events that could doom civilization: a nearby supernova explosion producing radiation that would scour the Solar System of life; the Sun's entry into a region of dense interstellar material; or the bombardment of the inner Solar System by millions of rogue comets from the Oort Cloud—the consequence of the close passage of another star.

Some people say, "Not to worry. Millions or billions of years is a very long time and we need not concern ourselves about the distant future." We say *stardust* to that, though we were really thinking of a stronger term! It is *never* too soon to start thinking about interstellar expansion for the long-term preservation of terrestrial life. The time is now!

After a Long Distance, Another Long Distance

Starflight is not just very hard, it is very, very, very hard! It is essential to really *feel* the fundamental cosmic distance scale we are up against. Say the Sun, about 1.4 million kilometers in diameter, is reduced to the size of a small marble with a diameter of 1 centimeter. On this scale, Earth is a barely visible dot about 0.1 millimeters in diameter, approximately one meter away from the marble Sun. The outer bounds of the Solar System—the orbit of Pluto—hover slightly beyond 42 meters on this scale. Another way of looking at it: the marble Sun could sit in the middle of a football field and the orbit of Pluto would then fit snugly between the two opposite goal posts!

Where on this scale is Proxima Centauri, which is the nearest known star beyond the Sun and is an actual distance of 4.3 light years away? On this scale, Proxima is about 292 kilometers removed, more than 80% of the distance from New York to Boston or from Washington to New York. Our present interstellar vehicles are literally much slower on this scale than crippled ants traveling between those cities. The *astronomical unit*, or AU, the average distance between Sun and Earth, is the relevant dimension to compare the realms of interplanetary and interstellar flight ($1 \text{ AU} \approx 149 \times 10^6$ kilometers). Proxima is 273,000 AU from the Sun. The piddling regime of interplanetary flight out to 40 AU is nearly 7,000 times less than this typical interstellar distance.

By the way, the small red dwarf, Proxima, is the third component of the triple star system, Alpha Centauri. The system's Sunlike "A" and "B" components revolve about a common center with a period of 80 years, separated by only 20 times the Earth–Sun distance. Proxima revolves around this pair far away with a period of millions of years.

Interplanetary flight, the kind of space travel we have done up until now, is thus about four powers of ten (four orders of magnitude or 10,000 times) less demanding in duration at any achievable cruise speed. Pioneers 10 and 11 leave the confines of the Solar System with an escape speed of about 2.5 AU/year, and Voyager 1 has already achieved and Voyager 2 will achieve (in August 1989 after encountering Neptune) a Solar System departure velocity of about 3.5 AU/year—that is, tens of thousands of years to over a hundred thousand years travel time if these spacecraft were on a direct trajectory to Proxima, which of course they are not (see Technical Note 1–1).

Because the velocity of a starship so dominates discussions of interstellar flight, it is important to consider units useful for measuring it. In the current era of chemically propelled rockets, units of kilometers per

The First Starships

The trajectories and fate of the first interstellar vehicles have been beautifully explored by Cesarone, Sergeyevsky, and Kerridge of the Jet Propulsion Laboratory, in a technical article that should be read to appreciate the reality of starflight *in the 20th century* (9). A summary of the authors' key projections:

	Pioneer 10	Pioneer 11	Voyager 1	Voyager 2
Launch Date	Mar. 3, 1972	Apr. 5, 1973	Aug. 20, 1977	Sept. 5, 1977
Loss of Signal	1994 (at 59 AU)	1996 (at 45 AU)	2012 (at 121 AU)	2013 (at 106 AU)
Departure Velocity Asymptotic (AU/yr)	2.4	2.2	3.5	3.4
Trajectory Angle to Earth Orbit Plane (degrees)	2.9	12.6	35.5	– 47.5
Closest Stellar Approach:				
Distance (ly)	3.27	1.65	1.64	0.80
Star	Ross 248	AC + 79 3888	AC + 79 3888	Sirius
Years to reach	32,600	42,400	40,300	497,000

Voyager 1 exceeded Pioneer 10's distance from the Sun in mid-1988 at 43 AU and subsequently will remain the most distant from the Sun of the four craft.

Voyager 2 exceeded Pioneer 11's separation from the Sun in early 1988 at 25 AU.

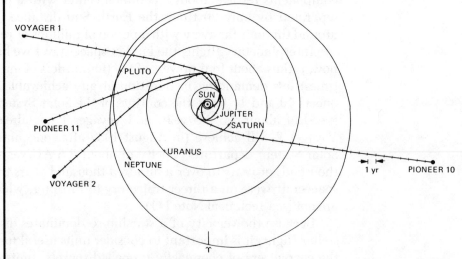

ECLIPTIC PLANE PROJECTION. PLANETS AND SPACECRAFT
POSITIONS SHOWN IN 2000 A.D.

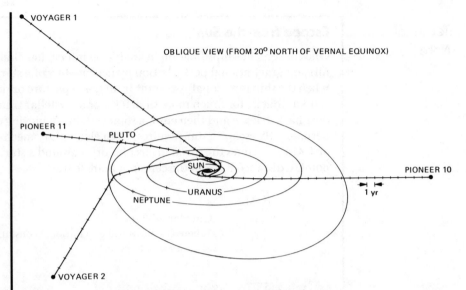

VOYAGER 1

OBLIQUE VIEW (FROM 20° NORTH OF VERNAL EQUINOX)

PIONEER 11

PLUTO

PIONEER 10

SUN

URANUS

1 yr

NEPTUNE

VOYAGER 2

Pioneer and Voyager trajectories. (Courtesy Jet Propulsion Laboratory, California Institute of Technology, Pasadena, CA, and JBIS)

second (km/sec) are quite suitable for describing spacecraft velocity. Moreover, km/sec is an appropriate unit for typical orbital speeds within the Solar System and even the velocity required to escape it. The speed of a spacecraft in a low Earth orbit is about 7.9 km/sec; the minimum velocity required to escape completely from Earth is 11.2 km/sec; Earth's orbital velocity about the Sun is 30 km/sec; and the minimum velocity required to escape the Solar System altogether starting 1 AU from the Sun is 42 km/sec (Technical Note 1–2).

Tens of kilometers per second are still insignificant compared to light speed, c, which is approximately 300,000 km/sec in free space. (The speed of light in vacuum is 299,792.458 km/sec.) Thirty km/sec, these days a luxurious pace, is only the fraction 10^{-4} c or 0.01% of light speed. Remember these often repeated facts: light sprints to the Moon from Earth in about 1.3 seconds and, if held to a circular path, light would in one second wrap 7.5 times around Earth's equator. The forementioned unit of AU/year might be useful in gauging the progress of some early precursor interstellar missions, but it is incompatible with decent starship velocities, considering that light speed is 63,500 AU/year.

By far the best measure of starship speed turns out to be simply the *fraction of light speed*, f_c, at which a spacecraft is traveling. What more

Escape from the Sun

One convenient simplification in analyzing interstellar flight is the lack of significant gravitational perturbation by nearby stars of a starship's velocity when the ship is reasonably distant from the departure or destination solar system, that is, for much more than 99% of interstellar transit. But a starship does have to leave and then enter a solar system. Achieving enough velocity to permanently escape a star requires, from elementary mechanics, achieving $\sqrt{2}$ or 1.414 . . . times the speed in circular orbit around a star at the departure point, a distance R_o from the center of a single star:

$$V_c = \sqrt{\frac{GM_*}{R_o}}$$

Circular orbital velocity
Where G = universal gravitational constant
M_* = mass of the star

$$V_e = \sqrt{2} \, V_c$$ Stellar escape velocity

These orbital and escape velocities, we have said, are generally *minute* compared with starflight cruise speeds necessary for reasonable transit times. Exceptions would be departure from circular orbit about a dense white dwarf star, or even more so, a neutron star. Any additional velocity beyond V_e that a starship achieves is referred to as its *hyperbolic excess velocity*, and is designated V_∞, read "V-infinity." For example, the hyperbolic excess velocities of the Pioneers and Voyagers are, respectively, 2.5 and 3.5 AU/yr. Once safely beyond the Solar System, they will cruise indefinitely at these speeds on trajectories that closely approximate straight lines.

appropriate standard to choose than the speed of the "craft" that do the fastest interstellar travel all the time—photons of light. It is much more convenient to speak of "0.1 c" than 30,000 km/sec or 635 AU/year. Moreover, f_c is easily converted to absolute terms by simply multiplying by about 300,000 to get speed in km/sec and by 63,500 for AU/year. But the really marvelous thing about the fraction of light speed unit is its relation to interstellar distances that are indelibly impressed on our psyches as light years. Despite the great utility to astronomers of the *parsec* ("parallax second," the distance from which 1.0 AU appears to subtend one arc second of angle [1 degree = 3600 arc sec.], the equivalent of 3.26 light years), the light year will forever be the unit of choice of the human starfarer, who reckons the cycles of life in years.

Light speed can be represented as 1.0 light year/year (ly/y), so whatever interstellar distance in light years is the objective, simply divide the light speed fraction, f_c , into it and get the time of the journey (at least as seen by home-based observers, i.e. neglecting relativistic time dilation for the star travelers). If a star is 15 ly removed and your speed is 0.1c, then it will require (15 ly)/(0.10 ly/y) or 150 years in transit (ignoring time to accelerate and decelerate in the case of a complete mission).

Scales of Distance

The boundaries of the Solar System's known planets fit within a sphere 50 AU in radius. What comes after that? First there is a vast belt, called the Oort Cloud, containing what are believed to be trillions of primordial icy comet nuclei, so sparsely spread through space that a starship departing the Solar System is unlikely to collide with a single one. This beehive of comets probably extends with decreasing density to a distance of 100,000 AU.

The nearest waypoint beyond the comet belt is the triple star system Alpha Centauri, at 4.3 ly, one of whose orbiting members, Proxima, happens to be the closest known star beyond the Solar System. Within a sphere of radius 21 ly, there are 75 known star systems (including the Sun) which contain 105 known stars, many of these being gravitationally linked double stars, with a few triples, quadruples, and even one quintuple system (see tables in Appendix 3 and illustrations in Chapter 2). Traveling beyond 21 light years, the number of stars in the expanding sphere of exploration increases approximately eight-fold with each subsequent doubling of the distance from the Sun. So within 40 ly there may be nearly 1000 stars, within 80 ly nearly 10,000 stars, and so forth. The formula can be extended only so far, however, because of the Sun's location out on an arm of a pancake-shaped spiral galaxy, the Milky Way, a whirl of stars that may contain between 400 billion and a trillion members. The Milky Way is approximately 70,000 ly in diameter, but is only a few thousand light years thick at the position of the Sun.

The spiral galaxy Andromeda is the major neighboring galaxy—at a distance of two million light years. Both Andromeda and the Milky Way, in turn, are members of a gathering of 20 galaxies called the Local Group, which stretch over millions of light years. Two minor galaxies, the Magellanic clouds—irregular aggregations of billions of stars—are closer to us than Andromeda. In fact, the spectacular supernova 1987 A exploded in one of these 160,000 years ago, and light from it has just arrived at the Sun.

The Local Group is gravitationally wedded to the Virgo supercluster, itself consisting of thousands of galaxies. Superclusters and even larger organizations of galaxies in the cosmos extend to the horizon of present visibility in the expanding universe—15 to 20 billion light years. The *visible* universe is defined by the age of the cosmos, an estimated 15 to 20 billion years. As the universe ages, this horizon grows farther away. Radiation from more distant parts of what may well be a virtually infinite cosmos (according to the new inflationary theory of cosmology) simply has not had time to get to us since the beginning of everything in the Big Bang "explosion." Hundreds of billions of galaxies populate the visible universe, but according to theory now seriously considered, even this visible universe may be a mere "atom" compared to the larger manifold of the inflationary cosmos. (See "The Self-Reproducing Universe," Eugene F. Mallove, *Sky & Telescope*, vol. 76, no. 3, September 1988, pp. 253–256.)

Of what significance to starflight is the architecture of this astonishing hierarchy? Its overwhelming size suggests, at least in the beginning, that we should be content to consider starflight out to a very limited range, perhaps to several tens or at most to one-hundred light years. There is ample interesting territory to explore within that realm. For neophytes, there is not much point in aiming at the stars beyond this zone before testing coastal waters. But someday. . . .

The Not-So-Fixed Stars

Though over the course of a human lifetime and even during long historical periods the stars may seem "fixed" on the celestial sphere, they are indeed moving with respect to one another. Referenced to the Sun, they have radial velocities toward or away from us that can be measured by the *Doppler* shifts of emission and absorption lines in their light spectra. They also have apparent *proper motion* across the sky, that is, movement tangential to the line of sight. So the stars are moving in three dimensions, and unless extremely fast starships are employed, it will be necessary to substantially "lead" a target star in order to rendezvous with it.

This is a consideration that will be taken up in greater detail in Chapter 12, but for now it suffices to note that typical stellar motions in the Sun's neighborhood are on the order of tens of kilometers per second. In other words, if starship velocities are of this same order (tens of km/sec, as they are now), the trajectory to a star from the Sun will be one side of a not very thin triangle (see Figure 1.1). The triangle would be

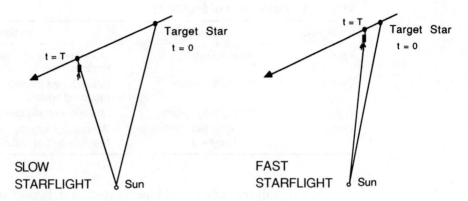

Figure 1.1 Aiming at a star.

very acute for extremely fast starflight. The other two sides of the big triangle are the initial Sun-target line and the star's approximately straight path through space to the rendezvous point.

Regimes of Starflight

Generating taxonomies of starflight will not get us one step closer to the stars, but it will help to put subsequent discussions in perspective. Above all, starflight has to do with patience—human patience, and human lifetimes. Without requisite patience and commitment, no one, no nation, and no world is going to try to cross the interstellar ocean. Therefore, the regimes of starflight are best characterized not by distances and velocities, but by transit times *as observed from the departure point*. At high velocity, the relativistic effect of time dilation makes shipboard time run slower, as the reader will no doubt have heard. (See Appendix 5.)

Our suggested categories of starflight are outlined in Table 1–1.

A few decades ago humanity was incapable of even Type-4 starflight. Now with the Pioneers and Voyagers we have embarked on a proof-of-principle version of Type-4. The once interplanetary craft will reach stellar distances by virtue of gravity whip momentum transfer in swings past the outer planets. But they will return no information other than the self-congratulatory "we did it"—if anyone on Earth is still around to check the calendar. Except for the possible dispatch of large human exploration colonies to the stars, *world ships*, humanity will probably leapfrog Type-4 starflight and before long embark on Type-3, with instrumented probes and perhaps even inhabited vessels.

Table 1–1 Categories of Starflight

Starflight Class	Transit Time	Rationale
Type 1:	10 to 100 years	Current human lifespan and planning horizon.
Type 2:	100 to 500 years	Enlightened extension of human planning horizon.
Type 3:	500 to 2,000 years	Major epochs in human history.
Type 4:	2,000 to 100,000 years and beyond	Beyond history, perhaps beyond the feasible global "attention span."

As we shall see, when and how Type-2 and Type-1 starflight will be accomplished gets considerably more speculative. The rationale for this schema is largely dependent on the average human *lifetime* and *lifespan* (the maximum age ever recorded at death) remaining what it is presently. However, even though contemporary lifespan is not far over 100 years, the explosive development of biomedical research holds great promise for major life extension, if we should choose to achieve it. The social consequences of a large increase in lifespan would of course be dramatic, pervasive, and possibly profoundly troubling. Perhaps one of the most predictable consequences, however, would be the more ready acceptance of extremely long-duration starflight. Thus, the problem of starflight is not only deeply entwined in human cultural perceptions, it is also inextricably tied to fundamental biology (see Chapter 14).

"Catch Me If You Can"

The potential long durations of early interstellar voyages create a glaring problem that has no parallel in human experience: A relatively slow vehicle dispatched too soon may be passed, long before it reaches its destination, by a more advanced technology, higher speed craft sent out much later. Would-be explorers of the New World may have been deterred by fears of sea monsters and falling off the edge of a flat Earth, but they did not hold back while anticipating a more efficient ride on the Queen Mary or hopping across the drink on the supersonic Concorde! Yet this is precisely the dilemma that may face initial voyages to the stars by people, and to a lesser degree by instrumented probes. The problem is one of setting out too soon: the "catch up" quandary.

In a way this is a very sticky subjective problem because it entails estimating technological and economic progress far into the future, a skill for which no good track record exists. However, Brice N. Cassenti considered a particular extrapolation of propulsion technology and concluded, "Missions on the order of 10 ly should be possible 200 years from now (4). Therefore, at this time, it appears that only propulsion systems capable of traversing distances at speeds that are a considerable fraction of the speed of light should be pursued, and only when these are shown to be infeasible should the Space Ark be considered." Curiously, physicist Freeman Dyson has arrived at the same time frame, albeit for different reasons. Dyson believes that we will launch large interstellar vehicles when annual GNP (gross national product)—or gross world product—grows to something like 1000 times its present level.

The "Space Ark" to which Cassenti refers is a self-contained world ship whose initial inhabitants would have long since died when their descendants reached the destination star system. The concept is venerable in the lore of starflight and science fiction, though as far as we are aware its earliest suggestion was published in 1929 by British crystallographer, J. Desmond Bernal (5). Since then, the idea has been elaborated in much greater detail by many other people, particularly in conjunction with efforts to establish orbiting space colonies within the Solar System (6,7). Generation ships, space arks, or world ships, have considerable merit for missions of interstellar colonization, and they will appear regularly in subsequent discussions.

Cassenti's and Dyson's conclusion seems to us reasonably secure, though we remain optimistic that unforeseen developments in propulsion technology will make feasible human missions to the stars beginning in the next two centuries. (In honor of the "catch me if you can" fable, perhaps we should name the first human interstellar mission Gingerbread Man–1.)

Starflight Propulsion

"Propulsion, propulsion, propulsion," might well be the interstellar explorer's equivalent of the real estate agent's exhortation of "location, location, location." Propulsion is the heart of interstellar transport, so before launching our tour through the "nuts and bolts" of starflight, behold the imposing array of candidate propulsion methods for attaining the stars:

• First, the classical generic self-contained rocket that gets both its energy and expellent mass completely from on-board reserves. Chemical rockets are all abysmally ineffective for starflight. Then there are the self-contained rocket's modern variants: the ion rocket (electric propulsion), which expels charged atoms (ions) at high velocity and perhaps uses a nuclear reactor to generate the requisite electric power; and the fission nuclear rocket: a nuclear reactor as power source to energize thermally accelerated hydrogen atoms (the reactor may be either a solid or liquid structure, or even gasified for maximum performance). Just over the horizon of technological feasibility is the fusion rocket, harnessing the energy of thermonuclear reactions to a high-speed particle exhaust. More far reaching still in their ultimate performance are varieties of anti-matter rockets, employing anti-matter/matter annihilation reactions as an energy source to accelerate and expel different kinds of particles and/or photons.

• Nuclear pulse propulsion resembles classical rocketry, as its energy supply and "propellant" are in the form of on-board micropellets of fusion fuel or rearward ejected bombs, which when detonated thrust the vehicle forward.

• Beamed power propulsion decouples the internal energy source of the classical rocket and puts it in either a laser, microwave, X-ray, or other kind of energy beam stationed within the Solar System. The beamed power can either energize propellant obtained from the interstellar medium or carried in the vehicle, to be expelled as in a rocket, or the beam can push the craft ahead by direct momentum transfer, thus eliminating the need for propellant.

• The interstellar ramjet is the analog of the terrestrial atmospheric ramjet. Interstellar space is a remarkably good vacuum, but using what little ambient matter exists in these barren reaches may make various types of interstellar ramjet feasible. A huge frontal area would be required to collect material in a large zone forward of the accelerating craft. Fusion reactions in the ingested material could create a high velocity exhaust jet.

• Interstellar solar sails. A close, high velocity pass near the Sun with a craft that unfurls a reflective sail could eject a payload from the Solar System at high velocity, using only the pressure of sunlight.

• Classical rockets, nuclear pulse, beamed power, interstellar ramjets, and advanced solar sails, though of primary interest, by no means exhaust the known possibilities of starflight propulsion (not to mention the ones yet to be conceived). An interstellar vehicle could accelerate by impulses received from impinging pellet streams beamed from the Solar System; it could increase velocity by traveling down a long linear electromagnetic launcher; or it could be propelled from a rotary momentum storage and transfer device.

• Speculative space propulsion. The sky is not the limit, if we can find ways to make use of undiscovered "loopholes" in physical law that may allow extremely fast transit between widely separated points in space-time.

• Combinations of many of the above. Often a symbiosis of systems has more interesting properties than any one by itself: for example, boosting an interstellar ramjet to high initial velocity using some form of classical rocket propulsion.

Ad Astra!

Completing this brief introduction to starflight, we remind skeptics that interstellar travel may, indeed, be virtually impossible, *if* their self-defeating assumptions are put in the way. Robert L. Forward, a pioneer inventor in the field of interstellar propulsion, has neatly outlined the artificial roadblocks typically set up by the nay sayers (8):

Stumbling Block 1. A starship must accelerate continuously at one earth gravity. Within a year, such a craft would reach about 0.77 c. (See the relativistic rocket equations in Technical Note 3–3, page 54.) But to continue accelerating at 1.0 gravity dictates greater and greater energy consumption—wasteful by orders of magnitude as the ship gets closer and closer to the speed of light, because the vehicle's mass increases from an effect mandated by relativity.

Stumbling Block 2. Interstellar travel must be performed with round trip times of only a few decades. First, a round trip is not the only kind of useful and interesting interstellar mission. But even if the vehicle must return, performing such a trip to one of the nearer stars over a minimum of *many* decades dramatically cuts the speed and energy requirement.

Stumbling Block 3. An interstellar vehicle must contain its entire reaction mass and energy supply on board. Nonsense! Make use of the abundant reserves of energy and matter in space. Think of beamed power and beamed mass. Consider sunlight and solar sails, and don't forget the interstellar ramjet.

If we at once admit the foolishness of these perennially suggested "impediments" to starflight, we will be well on our way to understanding that interstellar space does not need a bridge too far. Interstellar travel may still be in its infancy, but adulthood is fast approaching, and our descendants will someday see childhood's end.

2 Objectives of Interstellar Missions

West of these out to seas colder than
the Hebrides
I must go
Where the fleet of stars is anchored and
the young
Star-captains glow.

James Elroy Flecker, 1884–1915,
The Dying Patriot

The woods are lovely, dark and deep.
But I have promises to keep,
And miles to go before I sleep.

Robert Frost, 1923,
Stopping By Woods on a Snowy Evening

Types of Missions

Starflight comes in many flavors, but its two major genres are instrumented or automated missions and journeys by people. A few decades ago it was customary to refer to the former as "unmanned" missions and the latter as "manned" missions, but to avoid an admittedly slight gender bias it is better to call the latter missions "peopled" and the former "automated" or "robot probe" missions.

Usually, automated missions are conceived as expendable or one-way trips by robot equipment, though it is possible to contemplate round-trip automated probes, carefully biologically quarantined, that might return to the Solar System with samples of any biota found on extrasolar planets. (We say this cavalierly, but interstellar biological sampling may involve serious ethical issues. How "intelligent" must a species appear to be before we regard it as off limits to abduction?) In the class of one-way automated missions are "fly-through" probes that penetrate an extrasolar planetary system and collect data along the approach and exit trajectory.

Data gathering also would occur during the brief passes near the alien planets by the mother ship and the subprobes it would dispatch.

Next are one-way automated missions that decelerate at journey's end and go into orbit(s) around the destination star and its planets. The great advantage of this kind of mission is a prolonged period of observation and data taking—possibly lasting years, decades, or centuries—as well as the chance to dip into the atmospheres of extrasolar planets and perhaps land on their surfaces. The disadvantage of an orbiting/landing mission is the need to decelerate from interstellar cruise, thus putting a substantial additional requirement on the capability of the propulsion system to produce velocity changes. In determining the overall velocity requirement—referred to as "Delta-V" or ΔV—it must be remembered that mission phase velocities are *additive* for a complete journey. For example, add boost phase ΔV and deceleration phase ΔV to determine the propulsive velocity requirement for a one-way mission to orbit another star.

A much more demanding third and fourth class of automated missions are those that either fly through the destination solar system or orbit and land, but which then continue on toward other stars for further exploration (see Figure 2.1). For the present, the purpose of every category of automated mission is to collect and return scientific data, but there is another possible objective, one fraught with ethical questions and implications for the prevalence of life in the universe. It is possible to imagine missions that would "seed" extrasolar planets with terrestrial organisms or more advanced life forms. These *"directed" panspermia* missions would be the intelligently controlled analog of the theoretical and remote possibility that microbial life naturally drifts between the stars and survives to begin another evolutionary history on other worlds. Francis Crick and Leslie Orgel are among the most prominent investigators to have considered *directed* panspermia, though not the engineering technology required to bring it about (1).

Another conceivable purpose for automated starflight, one of its least plausible applications and one not to be attempted in the foreseeable future, are efforts to contact extraterrestrial civilizations by dispatching numerous "messenger probes" to dozens of extrasolar planetary systems. These *sentinel probes* would be long-duration watchtowers for the emergence of life and intelligence near promising stars. When and if signs of life emerged, the probes would relay their findings back to the Solar System. There has been considerable debate about how reasonable this approach would be for an advanced civilization—relative to the seemingly more "cost effective" electromagnetic signaling (4–6).

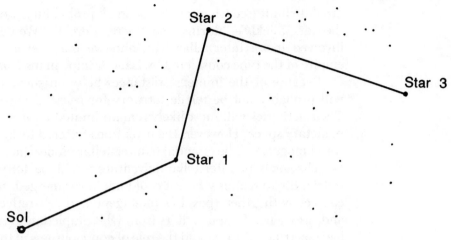

Figure 2.1 Multiple star mission.

Peopled missions are difficult because of long transit times even to the nearer stars. The most likely flights to be undertaken will be one-way trips to begin colonies in new solar systems. There could be round trips too, of course, but the question would be *who* would be coming back! Most likely the descendants of those who journeyed forth. For traveling to the stars *in person*, there are basically four generic alternatives:

1. *Biomedical extension of lifespan.* Allows traveling slow *or* fast in a world ship. If the crew lives long and happily enough, those who set out will arrive at planetfall.

2. *Low-speed worldship travel with current lifespan.* Generations would be born and pass on before reaching the target star system.

3. *Suspended animation or hypothermia/hibernation.* Allows starflight of indefinitely long duration.

4. *High velocity Type-1 starflight taking advantage of relativistic time dilation to shorten shipboard time.*

The Sociology of Starflight

Before considering much further what we believe the purposes of interstellar travel *will* be, we should state what they *will not* be. Unless all modern theoretical and experimental evidence is incorrect, and unless we find some unexpected loophole in physical theory, the speed of light will remain an absolute limit (see Chapter 13). So in all probability, we will never get to the stars quickly—"during a commercial break." We will likely never be involved in an interstellar war, and we may never create a galactic empire of the type considered by Isaac Asimov in his *Foundation Trilogy*.

Because of the immense distances between stars, interstellar space will probably not be fertile territory for celestial entrepreneurs either. Their activities will most likely remain limited to near-Earth and interplanetary space. However, the inventions fostered by Solar System space travel moguls will be essential to interstellar pioneering.

The goals of interstellar adventurers will be lofty: to spread terrestrial life to realms where it could not have emerged, to establish *direct* contact with other (possibly intelligent and self-reflective) life forms, and, as science fiction writers from Olaf Stapledon to Arthur C. Clarke have speculated, to expand the role of consciousness in the universe.

We suggest that one of two sociopolitical entities will provide the support for interstellar exploration and colonization. First, perhaps a single nation or consortium of nations might fund interstellar expeditions, perhaps embracing them as the "moral equivalent of war." Stonehenge, the Egyptian pyramids, medieval cathedrals, and the colonization of North America are prototypical long-duration programs that have challenged ambitious terrestrial societies. Another example, though with a shorter time-frame: the U.S./USSR sprint to the Moon in the 1960s.

We anticipate that national or supranational entities will someday be able to devote enormous discretionary wealth to an "interstellar initiative," many times the resources that are shamelessly consumed today in fratricidal military frenzy. Starships constructed by such benign nation states or consortia would not be severely constrained by cost and could therefore be as fast as technology allows.

An alternate initiator of interstellar human expansion might be the orbiting city-states proposed by Gerard K. O'Neill and others: self-contained space colonies that rotate to make use of artificial gravity. Initially serving the interests of Earth-based corporations and terrestrial nations, these world ships might become politically independent of Earth and spread throughout the Solar System, and their physical structure will be wrought from extraterrestrial resources—lunar, asteroidal, or cometary substances.

Some of the space colonies might decide to construct comparatively inexpensive "slow boats" and engage in flights to nearby stars lasting millennia. Other space cities, perhaps in the inner ranges of the Sun's comet belt, might hitchhike on perturbed comets and "diffuse" toward one of the nearby stars. Eric Jones and Benjamin Finney have considered in detail this possibility of ten-thousand year nomadic folk migrations (7).

In the twentieth century, Stone Age hunter-gatherers and nomads coexist with feudal societies and technologically advanced western and eastern communities. The same cultural diversity may well prevail in a future colonized Solar System. Therefore, different Solar System cultures may choose the options of fast ships, slow boats, or nomadic crossings. Nearby solar systems might be colonized many times by diverse groups of humans.

We expect our first fast ship or slow boat to be directed within the next few centuries to the possible planets of Alpha Centauri A or B, the Sun's nearest interstellar neighbors—discounting small Proxima Centauri, which probably cannot possess planets useful to humanity. Tens of thousands of years later, nomads from the Sun's comet belt might arrive in the cloud of comets that may surround Alpha Centauri.

In these voyages, different streams of humanity might experience direct physical contact only after tens of millennia of being separate. Isolated human subspecies possessing advanced technology might experience accelerated evolution and could move in directions that are now impossible to foresee. Would these human descendants recognize one another as being of common descent? Or would the relationship be more akin to contact between evolutionary lineages of completely independent origin, that is, true aliens? Physicist Freeman Dyson has written eloquently about this question in *Infinite in All Directions* (21):

When life spreads out and diversifies in the universe, adapting itself to a spectrum of environments far wider than any one planet can encompass, the human species will one day find itself faced with the most momentous choice that we have had to make since the days when our ancestors came down from the trees in Africa and left their cousins the chimpanzees behind. We will have to choose, either to remain one species united by a common bodily shape as well as by a common history, or to let ourselves diversify as the other species of plants and animals will diversify. Shall we forever be one people, or shall we be a million intelligent species exploring diverse ways of living in a million different places across the galaxy? This is the great question which will soon be upon us. Fortunately, it is not the responsibility of this generation to answer it.

There are, of course, still the doubters. The ordinarily imaginative astrophysicist Fred Hoyle made this astounding claim in 1983 (22): "Colonization of the galaxy is impossible, because it was *deliberately* arranged to be so." Hoyle was following in the footsteps of other believers in the "cosmic quarantine" hypothesis, such as John P. Wiley, Jr., who wrote in 1970, "With the rest of our solar system inhospitable to life as we know it and with travel to the stars problematical, man must lie in the bed he is making on Earth for the foreseeable future" (23). Astronomer Patrick Moore wrote in 1976, "I cannot believe that it will ever be feasible to send a manned space-ship out beyond the Solar System; my lack of faith in space-warps, time-warps, freezing techniques, and cosmical Noah's Arks is profound, though I am well aware that others do not agree" (24).

Yet there have also been notable converts from the ranks of the skeptical. The famous rocket pioneer Wernher von Braun wrote pessimistically in 1963, "With our present knowledge, we can respond to the challenge of stellar space flight solely with intellectual concepts and purely hypothetical analysis. Hardware solutions are still entirely beyond our reach and far, far away" (25). By 1969 von Braun had turned around completely, perhaps after seeing some of the spectacular conclusions about nuclear pulse propulsion that came out of Project Orion (see Chapter 4). He wrote, "The goals which have been identified for NASA in the years ahead do not go beyond the planets, but I am convinced that one day interstellar travel will become a reality" (26).

Today our perspective on crossing the vast interstellar gulf is dominated by the problem of propulsion engineering, so forgive our subsequent concentration on interstellar propulsion rather than on social futurology. Yet it is possible to outline a few scenarios in the foreseeable development of human culture and thus guide attention to interstellar propulsion concepts most likely to emerge in each instance (Technical Note 2–1). Starflight and contemporary myopic political organizations are incompatible, but because we will soon stand at the edge of the Solar System, it seems certain that Goddard's dream, as he wrote of it, "will not down" and will grow on us inexorably. We will take up Nature's gauntlet and not believe in "Nature's quarantine"—interstellar or otherwise.

Interstellar Precursor Missions

The Voyagers and Pioneers are not true interstellar precursor missions and were never intended to be, even though they will, indeed, be first to tiptoe into coastal interstellar waters. They lack staying power and the ability to return substantial data. What is needed are missions that will make a truly major plunge and travel perhaps 500 to 1000 AU from the Sun. Studies of such missions by the staff of the Jet Propulsion Laboratory (JPL) have already convinced the astronautical community of their near-term feasibility.

At a symposium, "Missions Beyond the Solar System," at JPL in August 1976, engineers and scientists considered the idea of a mission beyond the planets, yet not as far as another star, as a means to begin solving the engineering problems of eventual starflight. In November 1976 a NASA-funded precursor mission study began that produced a series of technical reports (8,9).

The study proposed a mission that would depart around the year 2000, last from 20 to 50 years, and return data from a distance of 400 to 1000 AU. Such an exploratory voyage has come to be called the "TAU (Thousand Astronomical Unit) Mission." Plans for year 2000 departure, though technically realistic if they had been financially supported, will now, of course, be delayed a decade or more. But the findings of the study are worth summarizing as a convincing example of starflight that could be done *now*.

Propulsion would be nuclear electric (NEP), basically a nuclear reactor or thermionic electric generation to power a low-thrust ion engine using perhaps mercury as fuel. The main scientific objectives would be to investigate the location and properties of the *heliopause:* the tenuous physical boundary between the Solar System and interstellar space, the interstellar medium, low-energy cosmic rays (presently excluded from our view by the heliopause), the mass of the Solar System (by trajectory analysis), and stellar distances by measuring optical parallaxes. The study group pointed out that the ship's data could be used to simulate, in "reverse time sequence," the approach of a starship to another star and its surroundings. Figure 2.2 depicts the TAU mission space probe.

Just as peopled expeditions to the nearby planets cannot occur before extensive automated reconnaissance, interstellar travel—even advanced robotic ventures—will depend on the findings of such a precursor mission. Looking to the future, the final report on the TAU mission also stated: "It is recommended that a subsequent study address the possibility of a star mission starting in 2025, 2050, or later, and the long lead-time technology developments that will be needed to permit this mission": *the first recorded semi-official recommendation to address true interstellar flight!*

A Techno-Social Futurology

The bizarre planet of *Homo sapiens sapiens* in the late twentieth century is at a major turning point as its human population pushes toward an inexorable doubling sometime in the next century, making further demands on already strained global resources. This is the reality in which dreams of starflight must incubate. A speculative look ahead at some hopeful as well as less optimistic possibilities:

Scenario 1. Nuclear disarmament within 25 to 50 Years; Earthlike worlds (ELWs) are discovered.

> In the decades following superpower disarmament, large worldwide programs would be needed to absorb the energies of technologists. One program could be starflight. If, for example, terrestrial or space telescopes reveal the presence of one or more Earthlike worlds (ELWs) circling the main suns of the Alpha Centauri system, starflight would receive a major impetus. Plans for automated probes would begin in earnest.

Scenario 2. Superpower competition continues for another century; Discovery of ELWs.

> Perhaps the "space-shield" bedecked superpowers would command orbiting laser banks in the 10^{10}–10^{11} watt range. Very low-mass, "electronic mesh" robots (see Chapter 5) could ride these laser beams on thin light sails at 10 to 20% light speed toward Alpha Centauri or other nearby stars.

Scenario 3. Slow human expansion into the Solar System; Alternating periods of international and interplanetary competition and cooperation; Development of space solar power satellites and independent self-sufficient solar system colonies; Discovery of many ELWs.

> Solar sail-launched robot probes requiring flight times of a few centuries might be used for initial interstellar exploration. This could be followed by solar-sail propelled human occupied "arks" requiring about a millennium to reach Alpha Centauri.

Scenario 4. Same as Scenario 3, but ELWs are not found.

> Civilization conducts proxy interstellar exploration using solar-sail launched robots. Solar-sail launched ships would simultaneously establish human settlements on the planets, moons, asteroids, and short period comets within our Solar System. Human interstellar expansion occurs only when other stars approach relatively close to the Sun (less than a light year). (When the Sun enters its red giant phase in about 5 billion years, solar-sail propelled ships become an efficient method of emigration to younger nearby stars.)

Scenario 5. Enduring world peace breaks out; Technological progress accelerates and leads to a fusion power renaissance; Radio signals are received from extraterrestrial civilizations.

Starships are unleashed from the bonds of current prosaic technology. Ample reasons have emerged for venturing far from the Solar System in fast, possibly relativistic starships. If the interstellar hydrogen fusion ramjet is feasible, it will become the propulsion system of choice. If not this, then ramjet derivatives such as the ramjet fuel-runway and the Ram-Augmented Interstellar Rocket (RAIR, see Chapters 7 and 8). Such craft offer the possibility of round-trip flights requiring mere decades in the reference frames of the starship crews. Direct contact between humans and extraterrestrials has potentially great import for the evolution of consciousness in the universe.

Scenario 6. An independent human civilization develops in the Sun's comet belt; ELWs are not found.

Solar-sail robots are used to begin interstellar exploration. Outward diffusion of colonies via comets or by planetary gravity assists might later place inhabited comets on slow interstellar trajectories.

Scenario 7. A unified Solar System civilization powered by sunlight develops.

The interstellar propulsion system of choice might be the solar-pumped laser or maser-driven sail. Two-way voyages would be possible, at least to the nearest stars, with round-trip flight times of about a century.

Scenario 8. Identical to Scenario 7, but the predominant power source is nuclear.

Initial interstellar expeditions are carried out with nuclear-electric propulsion or fusion propulsion—ramjet or rocket. Civilization might have the technological resources to "breed" large quantities of antimatter. Antimatter-propelled ships might be able to reach the nearer stars within a few decades. (At present, antimatter appears to be the most expensive route to starflight. Even a modest antimatter-propelled expedition would cost 100,000 current United States GNPs.)

Scenario 9. Global nuclear war occurs after extraterrestrial colonies are established, but before the outposts can operate independently.

Observing the ruined Earth, the survivors of Armageddon would be painfully aware of the fragility of life-bearing ecospheres in the cosmos. Although most of their resources would be devoted to survival, the still living would conceivably launch a few interstellar panspermia payloads as "genetic insurance," using solar sails or planetary gravity assist propulsion.

Scenario 10. Terminal global nuclear war occurs before off-planet colonies are established.

Humans and other mammals go the way of the dinosaurs, the dodo bird, and the woolly mammoth. Now mute, Pioneers 10 and 11 and Voyagers 1 and 2 cruise slowly through interstellar space, reminders to any spacefaring aliens encountering them by chance, of what once existed near Sol. The extraterrestrials use the maps aboard these probes to locate Earth and rummage in its ruins.

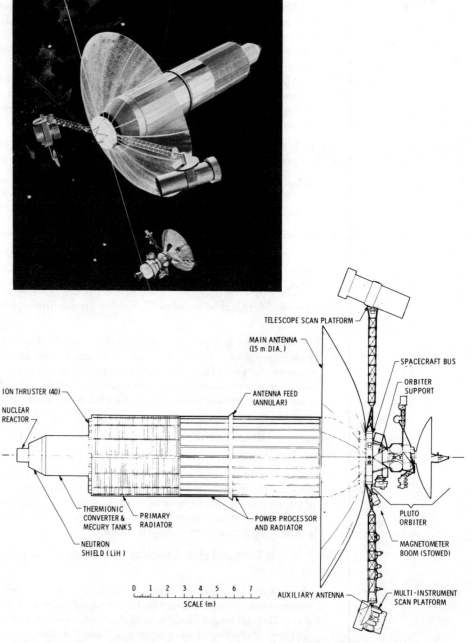

Figure 2.2 The TAU interstellar precursor probe. (Courtesy Jet Propulsion Laboratory, California Institute of Technology, Pasadena, CA)

Nearby Star Systems

The Sun lies within 100 ly of the geometric orbital plane of the spiral-armed Milky Way galaxy. The thickness of the spiral is some thousands of light years in our vicinity, and there are many millions of stars in a sphere around the Sun big enough to encompass the depth of the galactic disk. But the dawn of starflight cannot be that ambitious. For the moment, we should be satisfied with the domain out to perhaps 21 ly. This is a convenient measure, for within a sphere of that radius lies the nice round figure: 100 *known* stars contained within 75 star systems.

Another convenient property of the 21-ly zone: A starship encountering a solar system 21 ly removed on the day of birth of an earth-bound astronomer would radio back initial scientific data that will be newly received just in time for the starchild to analyze for her senior thesis in college!

We have synthesized a catalogue of known nearby stars from various compilations that have appeared in the literature (10–12). Appendix 3 lists the star systems and component stars (each often known by a few names and/or catalogue numbers) in order of increasing distance from the Sun. The table lists the star's direction in space, its location on the imaginary celestial sphere, in terms of (1950) right ascension and declination. These are coordinates most appropriate for the Earth-based observer but of little value to an interstellar navigator. He will be much happier with an x,y,z rectangular coordinate system, marked in units of light years *from the Sun:* hence the entries for "x," "y," and "z." Apologies for this anti-Copernican lapse, but starflight is one field in which, only relatively speaking, the Sun must serve as the center of our universe!

These special x, y, z coordinates, by the way, are still referenced to the Earth's polar axis pointing direction circa 1950 (Earth precesses like a top roughly once each 25,800 years.) The positive direction of the z axis points to celestial north (circa 1950); the x-axis is in the plane of Earth's equator (pointing in the direction of the vernal equinox); and the y-axis is perpendicular to the x-axis, forming a standard "right-hand" coordinate system. Figures 2.3, 2.4, and 2.5 portray the relative positions of the nearby stars, their numbers (in order of increasing distance from the Sun) being keyed to the data in Appendix 3.

We also list each star's tangential motion to the line of sight (its *proper motion*), measured in the usual units of arc-seconds/year across the celestial sphere, readily convertible to km/sec by multiplying this angle (in radians) by the distance to the star (in kilometers) and dividing by the number of seconds in one year. (Note also the specified *angle* of the

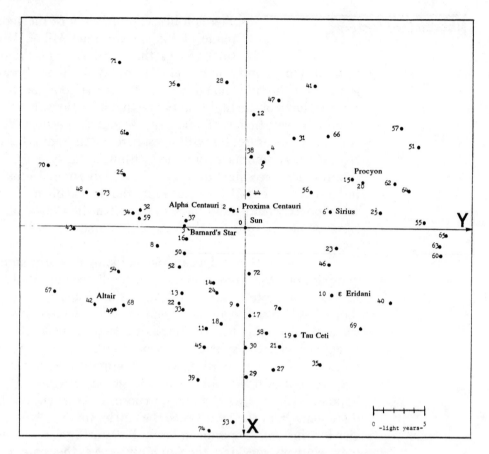

Figure 2.3 Positions of Nearby Stars in X-Y Plane. This is the plane of the celestial equator. Numbers refer to star systems listed in Appendix 3.

proper motion velocity vector from the direction of north, measured clockwise as seen from the Sun's perspective.) Each star's radial velocity (km/sec) is also listed: " + " meaning toward the Sun and " − " designating away from us. That completes the kinematic data on these nearby stars.

The most important physical parameters of the nearby stars are given in Appendix 3: approximate mass, luminosity, and *spectral type.* Not to digress too far into elementary astrophysics, but these properties are clearly the salient features of our stellar neighbors, apart from the much more critical question of whether they harbor planets. The known orbital characteristics of stars that are revolving gravitational companions of one another are not included.

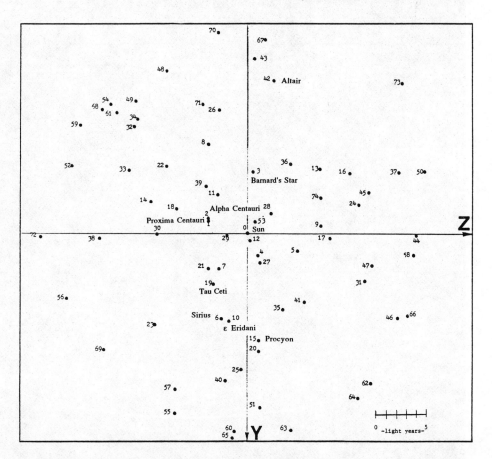

Figure 2.4 Positions of Nearby Stars in Y-Z Plane. The Y-Z plane is perpendicular to the plane of the celestial equator. Z-coordinates are positions above or below the plane of the celestial equator. Numbers refer to star systems listed in Appendix 3.

The most important factor in a star's evolution is the mass gathered into it from the contracting protostellar gas cloud. The mass determines the temperature of a stellar core when a star begins its thermonuclear conversion of hydrogen to helium. Running against the grain of intuition, a star that has more mass in the beginning does not live longer than one with less mass. The fusion burning rate of a more massive star is simply much greater. It takes a much shorter time to burn up a much larger amount of hydrogen fuel. For example, the 4.5-billion-year-old Sun will last another 5 billion years in a relatively stable hydrogen burning phase. But a star 10 times more massive than the Sun might

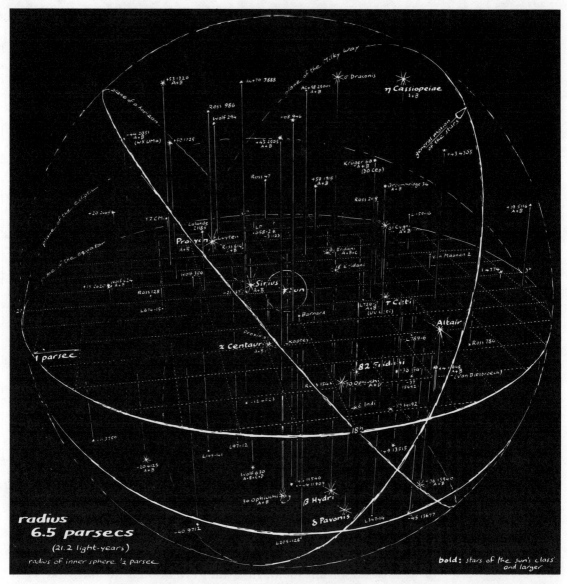

Figure 2.5 Nearby Stars in Three Dimensions. One parsec = 3.26 light years. (From *The Astronomical Companion*, Copyright 1979, by Guy Ottewell)

endure in a stable hydrogen burning phase for a mere 30 million years, while a star with only 0.1 solar mass could burn more than 3 trillion years!

The luminosity of stars with approximately the same initial composition and enrichment with metallic elements (which came from supernovae explosions elsewhere) rises strongly with increasing mass. A star's spectral type, on the other hand, is an indication of its surface tempera-

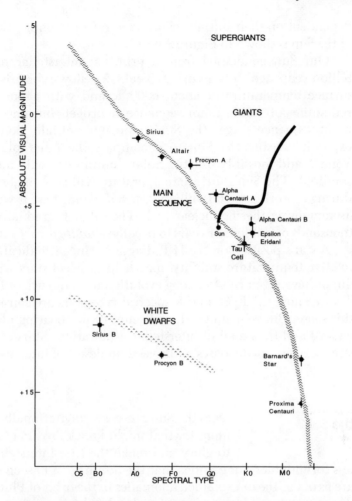

Figure 2.6 Hertzsprung-Russell (H-R) Diagram of Some Nearby Stars. The absolute visual magnitude and luminosity increases toward the top of the vertical scale—five magnitude units representing about 100-fold increase in luminosity. Surface temperature decreases in direction from O-type stars to M-type stars (e.g. G0, 6000°K; K0, 4000°K).

ture. The luminosity/spectral type information is conventionally graphed on what is called the Hertzsprung-Russell diagram or "H-R" diagram. Figure 2.6 is an H-R diagram of the nearby stars. Stars with roughly the same elemental composition as the Sun fall initially on a band called the "main sequence," which includes approximately 90% of all stars. As main sequence stars age and their thermonuclear burning characteristics change, they leave the main sequence band and follow contorted evolutionary paths on the H-R diagram that are strongly

LACOMBE COMPOSITE
HIGH SCHOOL LIBRARY

dependent on their initial stellar mass. As an example, the future course of the Sun is shown in Figure 2.6.

Our Sun condensed from a primeval interstellar nebula nearly 5 billion years ago. It is a fairly typical G2 yellow-white dwarf star with a surface temperature of about 6000°K and with about 5 billion years remaining of stable main sequence hydrogen burning. Following the main sequence stage, the Sun's core will initially contract and then expand, causing the Sun to encompass the inner planets, Mercury, Venus, and possibly Earth. Solar luminosity will increase a thousandfold. The Sun's surface temperature will cool to about 3000°K and during a profligate 100 million year *red giant* phase it will burn most of its remaining thermonuclear fuel. Then it will gradually contract over thousands of millions of years to become a *white dwarf* star.

A star's position on the H-R diagram, since it indicates the period of relative temperature stability, may help to select stars with planets that might have given rise to a long evolutionary sequence of life, at least life "as we know it." F, G, and K spectral-type stars are considered "best" in this sense. But remember, the existence of life-bearing planets is not the end-all and the be-all of interstellar exploration. Stars of all shapes and sizes may be worthy oases in the vacuum desert of interstellar space.

A Companion to the Sun?

Does the Sun possess a gravitationally bound companion star that might knock Proxima Centauri's claim to glory and make the first interstellar flight much less formidable? Suggestions that the Sun has an unseen orbiting cousin are as old as the unexplained anomalies in the orbit of Pluto that could be due to some co-orbiting planet or substellar body.

Since the early 1980s, the theory that the dinosaurs became extinct about 65 million years ago because of the impact of a 5 to 10 kilometer diameter asteroid has gained in scientific stature. Proponents of the theory have suggested that the sudden climatic change induced by the titanic impact of an intruding projectile could have wiped out the former lords of Earth (13). But when apparent periodicities in mass extinctions of other species were then noticed in the fossil record, a group of investigators began to suspect a celestial culprit of a different character.

They proposed that a dim dwarf star could be traveling around the Sun in a highly elliptical orbit (14–16). On the star's periodic swoops through the Oort comet belt (every 26 million years or so), it might gravitationally perturb many comets toward the inner Solar System and send a few dozen icy comet nuclei onto collision courses with Earth. The

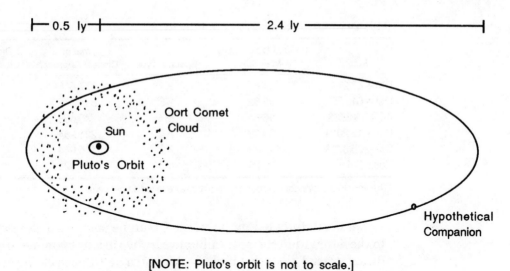

┠─ 0.5 ly ─┨──────────────── 2.4 ly ─────────────────┨

[NOTE: Pluto's orbit is not to scale.]

Figure 2.7 Hypothetical orbit of Nemesis. (Adapted courtesy Richard Muller)

theory caused a great stir in the scientific community in the early 1980s, and the question is still far from resolved. The companion was even dubbed Nemesis—for obvious reasons. But no search of the heavens for the suspected solar companion has succeeded in finding the "death star," the supposed master of the fates of so many millions.

If Nemesis exists, its orbit is estimated to have a semi-major axis on the order of 150,000 AU, and it would periodically come within 30,000 AU of the Sun (Figure 2.7). Even though Nemesis or its equivalent might someday turn up in a search, it seems more likely that the difficulty of interstellar flight will not be changed by finding such a nice intermediate "way station." For one, Nemesis at best might cut the shortest interstellar path by a factor of two to ten and we would have to wait millions of years for the best opportunity to leap across to the other star. Though the Moon has served as a destination to whet our appetites for the more distant planets, there is likely to be no "Moon" on the way to Centauri.

Other stars may occasionally approach much closer to the Sun than Alpha or Proxima Centauri. Some of these close approaches may gravitationally disrupt comets in the Sun's Oort comet belt, directing them sunward as immense comet showers. One of these bombardments of the inner solar system may have contributed to the demise of the dinosaurs about 65 million years ago, even if an *orbiting* Nemesis star does not exist.

As part of the trajectory analysis for the interstellar phases of the Voyager 1 and 2 and Pioneer 10 and 11 missions, a NASA Jet Propulsion

Table 2-1   Close Stellar Encounters

Star	Present Distance (ly) from Sun	Spectral Type	Year of Closest Approach	Distance (ly) of Closest Approach
DM + 62 274	44.69	K1	477,816	1.61*
DM + 61 366	32.62	K5	814,872	0.29*
AC + 79 3888	16.64	M4	40,598	2.96
DM + 45 2014	39.30	K4	221,964	1.54
DM + 25 3719	43.49	K2	175,944	1.66
Ross 248	10.26	M6	34,923	2.90

*The error in the closest approach distance may exceed 1 light year.

Laboratory study led by Robert J. Cesarone computed close approaches to the Sun within the next million years by various known main sequence dwarf stars. The study (Table 2–1) found a half-dozen close approaches.

Some astronomers are currently searching for *"brown dwarfs,"* objects that are intermediate between small stars and large planets, having 0.01 to 0.085 solar masses. If brown dwarfs turn out to exist, some might be much closer to the Sun than even these near passes. Astronomers have also conducted numerous searches for what some have presumed to be a tenth planet of the Solar System beyond the orbits of Pluto and Neptune: photographic hunts predicated on still unexplained features of Pluto's orbit. Nice as it would be to have "Planet X" turn out to be a tiny brown dwarf star, it seems unlikely that this will be the case.

Picking a Destination

Content with what we already know of the solar neighborhood, what targets are most opportune? That depends on the mission objective. If it is to colonize a planetary system, it would be best to pick a star with planets or at least one not likely to change its luminosity erratically or too soon. If hunting for indigenous extraterrestrial life is the agenda, then a Sun-like star with planets should also be our pick. It might be better, however, to have the very wide-ranging objective to explore varied astrophysical terrain and so consider all the nearby stars as candidates.

Others have made "wish lists" of nearby stars (17–20), which we think are great fun. Nevertheless, they do seem like lists prepared by beggars deciding which mansions they would prefer to live in when they become rich. Simply, the nearer the better, so the Alpha Centauri system and Barnard's star rank highest. But following in the traditional path of earlier "beggars," we note in Table 2–2 a few other interesting way stations in the void.

Table 2-2 Nearby Star Destinations of Choice

Initial Voyages by Robot Explorers

"Proof of Principle" Missions

Alpha Centauri (4.3 ly):

Alpha Centauri A: 1.11 solar mass, a yellow spectral class G2 (identical to Sun) star.

Alpha Centauri B: 0.85 solar mass, an orange K-1 star (23 AU from component A).

Alpha Centauri C (Proxima): 0.10 solar mass, a young red M-class dwarf, an irregular flare star, orbits A and B pair at 12,000 AU. Thought to have been captured by A/B pair within the last few billion years.

Barnard's Star (5.9 ly):

Mass: 0.15 solar, M3 class red dwarf, thought by astronomer Peter van de Kamp from his astrometric observations to have two planets (disputed by others).

Exobiology

Alpha Centauri (4.3 ly)

Component A more likely than B to have life-bearing planets, but 23 AU separation of A and B possibly not conducive to formation of planets or evolution of life.

Epsilon Eridani (10.8 ly)

0.75 solar mass, K2 spectral type, the star closest to the Sun that is most like the Sun and not part of a multi-component star system. A young star, possibly not old enough for life to have evolved to an advanced level. One of two stars examined by Frank Drake in first SETI search—Project Ozma, 1960.

Tau Ceti (11.8 ly)

0.9 solar mass, G8 spectral type. Star most like the Sun, but cooler and not as bright. Other star in Drake SETI search in 1960.

Exotic astrophysics

Sirius (8.7 ly)

Sirius A: 2.2 solar masses, hot A-1-type star.

Sirius B: A hot white dwarf star, about one solar mass crammed into an Earth-size sphere. During its lifetime may have contributed some of its mass to Sirius A.

Colonization Missions

Alpha Centauri
Tau Ceti
Epsilon Eridani

3 Rocket Propulsion for Interstellar Flight

Climb high
Climb far
Your goal the sky
Your aim the star.

Inscription on Hopkins Memorial steps,
Williams College, Williamstown, Massachusetts

Rocket Fundamentals

So entrenched is the rocket in today's space flight that it is still considered mildly outrageous to suggest that any other means might be viable for space transport. Solar sails, laser or microwave pushed sails, and interstellar ramjets work on principles fundamentally different from the rocket, and these nonrockets may ultimately be the first road to the stars. But being so close to the earliest days of astronautics, we naturally reserve for the rocket a special place in our minds and hearts. The rocket is the starting point in thinking about methods of reaching the stars.

A rocket is distinguished from other kinds of propulsion by being totally self-contained. The classical rocket receives no mass or energy from the outside environment. It works exclusively on the principle of conservation of momentum, which manifests itself in Newton's well-known Third Law concerning equal and opposite forces of action and reaction. Nuclear pulse propulsion, also completely self-contained, in principle is akin to classical rocketry. But the blasting of a pusher-plate by a rearward-ejected bomblet or the rocketlike roar of repetitive fusion microexplosions differs enough to merit its own category, hence the next chapter.

The one-stage classical rocket is so simple that it is easy to forget how elegant it is. It does not, of course, "push against" the outside world: the common misconception of the uninitiated. (At the time of his early experiments, Robert H. Goddard was momentarily given a bad name in a notorious *New York Times* editorial contesting this fact!) In fact, a rocket works best in the vacuum of space rather than in the atmosphere of a planet: (1) Because its high-speed exhaust meets no resistance and is not slowed down, and (2) No gas-dynamic drag retards the vehicle's forward motion. The basic principle of rocket propulsion is portrayed in Technical Note 3–1. From this it is seen that the higher the speed of the rocket exhaust, V_e, the less initial mass (mostly propellant) is required to boost the payload and remaining structure to a particular final velocity.

This gives rise to the extremely important concept of *specific impulse* or "I_{sp}", the common measure of all rockets. Specific impulse is a measure of the efficiency of a rocket: how much impulse (thrust multiplied by time) is produced per unit of mass of propellant expenditure. Specific impulse turns out to be equal to exhaust velocity, but by convention it is always given in units of seconds. This requires dividing exhaust velocity by the constant, g, the acceleration of gravity at the Earth's surface (9.81 m/sec):

$$I_{sp} = \frac{V_e}{g}$$

If given I_{sp}, to calculate V_e approximately in meters/sec, simply multiply I_{sp} by 10, since g is so close to exactly 10.

A rocket obviously can attain a final velocity greater than its exhaust velocity: how much more so depends on how much mass it has at ignition versus when its fuel has been used up. Remember that exhaust velocity is reckoned as the velocity of the exhaust stream *relative to the rocket*, not to the Earth, the Solar System, or to some other reference frame. However, the rocket's fundamental problem is that so much energy needs to be expended just to boost a large amount of propellant to a certain velocity so that that propellant, in turn, can give the remaining propellant and structure an incrementally higher velocity. Moreover, the requirement for propellant mass does not simply increase proportionally to the final velocity ΔV that the vehicle must achieve, but *exponentially*! Witness the *rocket equation* from the derivation in Technical Note 3–1 (M_o is the initial mass of the vehicle and M_f the mass at "burnout"):

$$\frac{M_o}{M_f} = e^{\frac{\Delta V}{V_e}}$$

It is possible to build rocket structures that have only several percent structural mass, including engines and tankage. In other words, most of the rocket takeoff mass consists of propellant. However, when the final velocity required for a mission is large, it becomes exceedingly difficult

Basic Principle of Rocket Propulsion

By conservation of momentum, the small change in velocity (dV) produced in a rocket of mass, M, by a small quantity of mass (dM) exiting the rocket engine with exhaust velocity, $-V_e$, must keep the total momentum of the system (rocket plus exhaust) zero, which is what the total momentum is initially:

$$MdV - V_e dM = 0$$

Rearranging:

$$dV = V_e \frac{dM}{M}$$

Integrating:

$$\int dV = V_e \int \frac{dM}{M}$$

The result of the integration may be expressed as:

$$\Delta V = V_e \ln \frac{M_o}{M_f}$$

That is, the total velocity increment, ΔV, imparted to a rocket with exhaust velocity, V_e, is V_e times the natural logarithm of the mass ratio—the initial mass divided by the final mass. This is called the *rocket equation*. Unfortunately, the natural logarithm function increases very slowly with its argument, M_o/M_f, so even a large mass ratio does not lead to hefty multiples of V_e:

M_o/M_f	ΔV (in multiples of V_e)
10	2.30
100	4.60
1000	6.91, etc

An alternate way of expressing the rocket equation:

$$\frac{M_o}{M_f} = e^{\frac{\Delta V}{V_e}}$$

to design a single stage with a structural fraction small enough to achieve the required ratio, M_o/M_f. The principle of the staged rocket enters to the rescue.

The Staging Principle

One of the most important innovations in rocket technology is the use of multiple rocket stages to achieve higher final velocity for the payload. By dropping off parts of the rocket along the way, less total mass needs to be accelerated by the engines of succeeding stages. If a single rocket stage can boost the remaining vehicle (temporarily considered as "payload") to final velocity, V_1, then a properly proportioned two-stage rocket can achieve $2\,V_1$, a three-stage rocket $3\,V_1$, and so on. The key phrase is "properly proportioned," with each stage having the same exhaust velocity. As each stage is separated, the remaining rocket structure—considered as payload for the stage immediately preceding it—must have a ratio of "payload" to initial mass equal to that ratio for the preceding stage. This is best seen diagrammatically in Technical Note 3–2.

It is not essential that the stages be proportioned this way, and indeed it is not easy to build rockets with such proportions, but it has been shown that this is the most effective staging approach, if each stage has the same engine exhaust velocity (specific impulse).

Conventional Rocket Technology

Rockets that rely on chemical energy have a long history that probably can be traced to the Chinese civilization of the twelfth century A.D. We even have the chemical "rocket's red glare" enshrined in the U.S. national anthem, testimony to the bombardment of Fort McHenry in the War of 1812 by British Congreve rockets.

Solid propellant rockets have evolved from the lineage of fireworks and include the two solid-rocket boosters of the U.S. space shuttle and most U.S. ballistic missiles. Solid chemical propellants are mixtures of *fuel* and *oxidizer*, which when ignited at the surface of a cast propellant "grain" liberate tremendous thermal energy that is then channeled into directed motion via a convergent-divergent (so-called de Laval) supersonic exhaust nozzle (Figure 3.1). The energy released in solid propellants (and all other chemical rocket propellants) derives exclusively from the sudden rearrangements of electrons in their shells surrounding

The Staging Principle

For a rocket proportioned as in the diagram and with equal exhaust velocities and structural factors for each stage, the velocity achievable by n stages is:

$$\Delta V = n[V_e \ln(\mathbf{R})]$$

where $\mathbf{R}$ represents the initial-to-final mass ratio of the assemblage at each staging (identical in each case). This kind of staging multiplies by n the final velocity achievable by a single stage rocket of mass ratio $\mathbf{R}$, but with the penalty of an enormous *overall* mass ratio. The overall mass ratio of the vehicle becomes:

$$\frac{\text{Total initial mass}}{\text{Burnout mass}} = \mathbf{R}^n$$

Note that *irrespective of the number of stages*, n:

$$\mathbf{R}^n = e^{\frac{\Delta V}{V_e}}$$

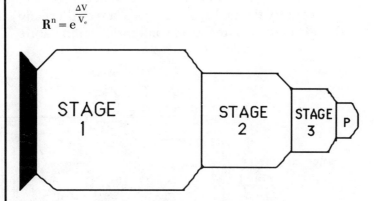

Schematic of the rocket staging principle.

atomic nuclei. In chemical propulsion, the nuclei merely go along for the ride and are unchanged.

Liquid propellant rockets employ chemicals in liquid form and are generally more complex because of the need to pressurize and inject a fine spray of propellant droplets into a combustion chamber of high temperature and pressure gas (Figure 3.2). Among his many other contributions to astronautics, Robert Goddard has the honor of having launched the world's first liquid propellant rocket from his aunt's farm

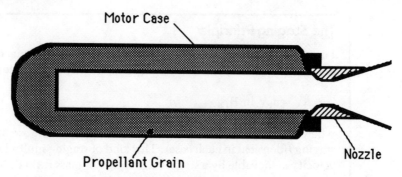

Motor Case

Nozzle

Propellant Grain

Figure 3.1   Solid propellant rocket.

near Worcester, Massachusetts. Liquid propellant rockets may be mono-propellant or bipropellant, but bipropellant combinations generally produce much more energetic reactions and higher velocity exhausts. (Some exotic tripropellant combinations have also been suggested.) Excellent treatments of chemical rocket systems are found in References 1 and 2.

Worth noting is that the I_{sp} of chemical rockets (and all other rockets that convert thermal motion into a directed exhaust) increases proportionally to the square root of combustion temperature, T_c. Increased I_{sp} is

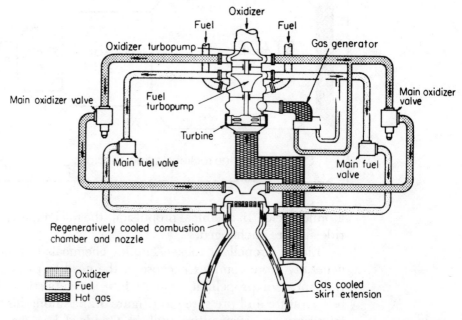

Figure 3.2   Liquid propellant rocket. A schematic of a typical propellant flow system and cooled nozzle. (Courtesy NASA)

Table 3-1 Specific Impulse of Advanced Chemical Propellants (57)

Propellant Combination	Specific Impulse (seconds)
Hydrogen-Fluorine (F_2/H_2)—ideal	528
Hydrogen-Oxygen (O_2/H_2)—space shuttle	460
O_2/H_2 (ideal)	528
O_3/H_2	607
$F_2/Li\text{-}H_2$	703
$O_2/Be\text{-}H_2$	705
Unproved Exotic Chemical Concepts	
Free Radicals $(H + H) \rightarrow H_2$	2,130
Metastable Atoms (e.g. Helium)	3,150

also caused by a lowering in the average molecular weight, MW, of the chemical exhaust products as seen in the equation ($\propto$ means "proportional to"):

$$I_{sp} \propto \sqrt{\frac{T_c}{MW}}$$

Chemical rockets are also characterized by high thrust-to-weight ratios and are thus ideal for lifting off a planetary surface. Other kinds of rockets, such as electric thrusters or fusion rockets have thrust/weight ratios much smaller than 1.0.

Where does chemical rocketry stand in the competitive world of I_{sp}? Table 3-1 shows that the growth potential of chemical rocketry is, indeed, limited. Fortunately, we will encounter many other kinds of rockets with I_{sp} much greater than chemical rockets. The specific impulse of a rocket with the highest exhaust velocity possible—the speed of light—would be about 30,000,000 seconds.

Electric Propulsion

Electric propulsion is a practical way to achieve very high specific impulse with current technology (6).

Electric or "ion engines" have been built, tested, and flown, though not on missions beyond Earth orbit. Electric propulsion basically sacrifices high thrust for high exhaust velocity: no problem if the rocket is already in orbit and thus never needs a thrust/weight ratio greater than 1.0.

The principle of the ion engine is illustrated in Figure 3.3. An electric power generator supplies energy to ionize (charge) the atoms of the propellant and also provides the power for electric fields to accelerate these ions to high directed velocity. The much lighter electrons that are stripped-off in the ionization process are sent rearward to combine with the ion exhaust and neutralize the charge in the beam (otherwise the rocket would build up an intolerable negative electric charge). Liquid mercury is an example of the kind of propellant used in ion engines: any element that can be ionized and accelerated efficiently. The electric power generator might be, for example, a nuclear reactor, or an array of photovoltaic cells that captures sunlight or laser light.

The specific impulse of an ion engine is most generally:

$$I_{sp} = \frac{1}{g}\sqrt{2\frac{q}{m}V_a}$$

where q and m are the charge and mass of an individual ion and V_a is the voltage or potential difference through which the ions are accelerated. Ion engines have been built with specific impulse in the range 2,500 to 10,000 seconds. The furthest this performance might be extended, though perhaps a wild extrapolation, is to 400,000 seconds (7,8).

Nuclear Rockets

Nuclear rockets based on fission or fusion ideally could far exceed the limitations of rockets using chemical energy. But even though for fission the potentially usable energy content per unit mass of fuel is 5.5×10^6 times greater than chemical energy and for fusion it is 2.6×10^7 greater, practical designs may severely limit the prospects of fission and fusion rockets. In a chemical reaction, an insignificant amount of mass is converted to energy as electrons rearrange themselves among the electron clouds of combining atoms. In uranium fission, a significant fraction of the nuclear fuel mass, ϵ, is *potentially* convertible to directed energy of the exhaust, namely 7.9×10^{-4}, which represents complete fuel fissioning or "burnup." (In practice, contemporary nuclear reactors fission only about 1% of the atoms in their fuel, but 7.9×10^{-4} is the fraction of the fissioned mass converted to energy.) In fusion rockets based on deuterium reactions, $\epsilon = 4 \times 10^{-3}$.

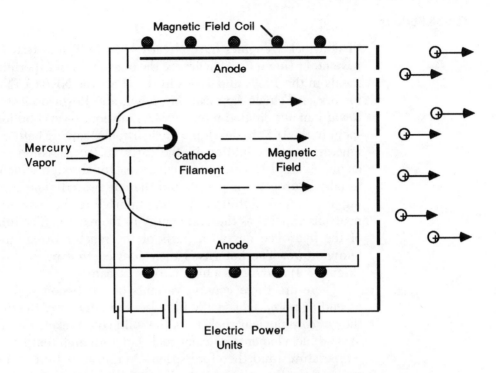

Magnetic Field Coil

Anode

Mercury
Vapor

Cathode
Filament

Magnetic
Field

Anode

Electric Power
Units

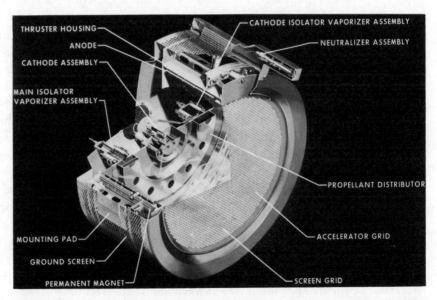

THRUSTER HOUSING

ANODE

CATHODE ASSEMBLY

MAIN ISOLATOR
VAPORIZER ASSEMBLY

CATHODE ISOLATOR VAPORIZER ASSEMBLY

NEUTRALIZER ASSEMBLY

PROPELLANT DISTRIBUTOR

ACCELERATOR GRID

MOUNTING PAD

GROUND SCREEN

PERMANENT MAGNET

SCREEN GRID

Figure 3.3 Ion rocket engine. (Photograph courtesy John Wiley & Sons)

Fission Rockets

Fission rocket engines have already been built and tested, though they have never flown (9). They were the outgrowth of experimental developments in the 1950s and 1960s in the U.S.: the NERVA (Nuclear Energy for Rocket Vehicle Application), Kiwi, and Rover projects. With a view toward an anticipated manned Mars mission, several billion dollars were spent in the U.S. to develop a nuclear rocket engine before the program's ignominious termination in the early 1970s. A fission-powered rocket engine is simply a compact nuclear reactor through which propellant—usually hydrogen—is passed and thereby heated. The energy of fissioning uranium or plutonium nuclei imparted to the fission fragments and neutrons appears as thermal energy in the reactor. The intimate contact of the hydrogen gas with parts of the reactor brings the gas to high temperature. The hot gas is then allowed to expand through a nozzle (Figure 3.4) and forms a high velocity stream.

There are three generic variants of the fission rocket, in order of ascending potential specific impulse: the *solid core*, the *liquid core*, and the *gas core* nuclear rocket. In the solid core rocket, the reactor consists of solid fuel elements: nuclear fuel clad with high temperature alloys. Its temperature (and therefore specific impulse) is limited by the need to preserve the integrity of the solid fuel elements, that is, to prevent them from melting. The liquid-core and gas-core rockets for which many, many mechanical designs have been conceived could in theory go to much higher temperatures because they relax the requirement to keep the nuclear fuel solid. In a typical liquid-core design, hydrogen gas is forced through a spinning annulus of microscopic liquid nuclear fuel droplets, the purpose of spinning being to retain the nuclear fuel and sustain the fission reaction. Climbing to a still higher temperature, the gas-core fission rocket spins gaseous nuclear fuel in a high temperature vortex to permit the loss of as little nuclear fuel as possible.

The specific impulse ranges for fission nuclear rockets are: solid core (500–1100 sec), liquid core (1300–1600 sec), and gas core (3000–7000 sec). Nuclear rockets achieve their high specific impulse in part because of the low average molecular weight of their exhaust products (see earlier equation), that is, diatomic hydrogen gas that has been significantly broken down into individual hydrogen atoms and ions. Fission rockets, like their chemical cousins, have high thrust/weight ratios, but given their inherent much greater energy content, their specific impulse falls far short of what might have been expected.

SOLID-CORE NUCLEAR ROCKET

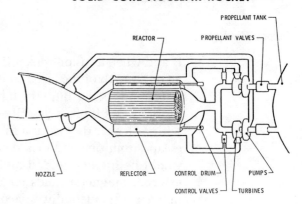

LIQUID-CORE NUCLEAR ROCKET

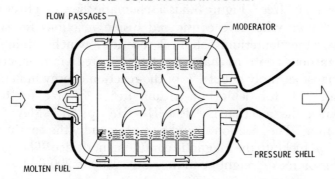

OPEN-CYCLE GAS-CORE NUCLEAR ROCKET

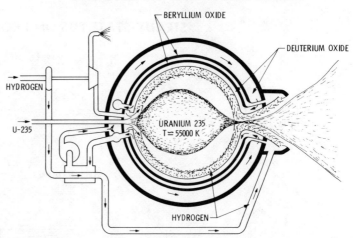

Figure 3.4 Nuclear rockets: solid, liquid, gas core. (Courtesy U.S. Government)

Fusion Rockets

If and when a fusion rocket is built, it is likely to be a massive device and an outgrowth of the large controlled fusion experiments for power generation now being built in laboratories around the world (13–33). The fusion rocket is often described as a magnetic plasma confinement bottle with a "leak"; that is, in a fusion rocket hot plasma would be allowed to escape from one end of a controlled fusion reactor. Perhaps the plasma would be further guided and accelerated into a focused exhaust jet by external magnetic fields created by high-temperature superconducting magnets. Incidentally, space is the natural place for a fusion reactor because there the high vacuum required to sustain a thermonuclear plasma is "free." Figure 3.5 is a schematic view of a fusion rocket engine.

Those who have considered fusion reactions for space propulsion favor the deuterium/helium-3 reaction, which minimizes the flux of hazardous neutrons that cannot be directed via magnetic fields into an exhaust jet (only about 1% of the reaction energy in neutrons compared with 75% for deuterium-tritium fusion favored for terrestrial power plants; the reaction is $^3He + {}^2H \rightarrow {}^4He + p$; 14.7 Mev proton, 3.6 Mev helium). The reaction would be sustained by the continuous injection of fusion fuel into the magnetic confinement device. The specific impulse of a fusion rocket might be in the range 2500 to 200,000 seconds, though its thrust/mass ratio should be expected to be only between 10^{-4} and 10^{-5}.

STEADY-STATE FUSION PROPULSION

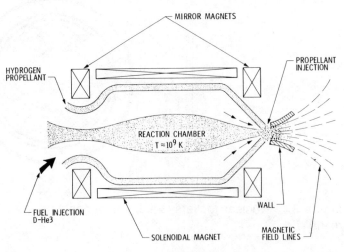

Figure 3.5 Plasma fusion rocket. (Courtesy U.S. Government)

Antimatter Rockets

If a rocket could have an exhaust velocity equal to the speed of light, about 3×10^8 m/sec, it would have the largest possible specific impulse, namely about 3×10^7 seconds. Perhaps lured by that prospect, the German rocket scientist, Eugen Sänger, in the early 1950s (34) conceived the antimatter rocket or "photon rocket," as he called it. His idea was to use the mutual annihilation of matter and antimatter to produce energetic gamma rays, radiation that like all other photons of electromagnetic radiation (including visible light, radio waves, X-rays, etc.) travels at the speed of light.

The only kind of antimatter known in Sänger's day was the positron, the positively charged electron, so he imagined a rocket in which stored positrons would be metered out to annihilate electrons and thus create an intense beam of energetic gamma rays with wavelengths about 10^{-5} that of visible light. Sänger could not surmount one problem, however, which proved the undoing of his antimatter rocket design. The two gamma rays produced in a positron-electron annihilation come out in random directions. To make a rocket, rather than a "gamma ray bomb," a way had to be found to channel the gamma rays into a directed exhaust stream.

Sänger tinkered with the idea of using an "electron gas" as a mirror to reflect and channel the gamma rays, but nothing he designed seemed to work. At the moment, no one in the advanced propulsion community spends much time thinking about the Sänger antimatter rocket (Figure 3.6). Fortunately, a way has been found around the problem: use of a different kind of antimatter, namely antiprotons. Antiprotons and protons mutually annihilate and produce short-lived elementary particles called *pi-mesons* or simply, *pions:* an average of 1.5 positively charged pions, 1.5 negatively charged pions, and 2 neutral pions per proton-

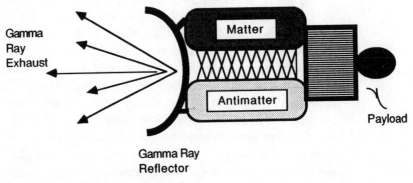

Gamma
Ray
Exhaust

Matter

Antimatter

Payload

Gamma Ray
Reflector

Figure 3.6　Sänger's photon rocket.

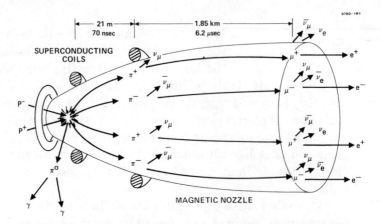

Figure 3.7   Antimatter rocket. (Courtesy Robert L. Forward)

antiproton annihilation. The neutral pions decay quickly to gamma rays, but the charged pions wait longer to decay—ultimately into gamma rays and neutrinos.

Physicist Robert L. Forward and others (36–39) have proposed using a "magnetic nozzle" generated by superconducting coils to channel the charged pions of proton-antiproton reactions into a directed exhaust (Figure 3.7). By the time most of the gamma rays appear in the exhaust, their point of origin is tens of meters behind the rocket. Meanwhile, the energy of the charged pions has been channeled into the momentum stream of the exhaust. Alternatively, the pions could be used to heat a larger mass of inert propellant, such as hydrogen, water, or methane, which could then be channeled into a directed exhaust. Forward suggests that 30 to 50% of the annihilation energy could be transferred to directed exhaust. A few milligrams of antimatter could replace the energy in tens of tons of chemical propellant.

So what we have now is definitely an antimatter rocket but certainly no longer the photon rocket that was imagined in earlier days. That fantastic specific impulse of 3×10^7 will not be realizable. But the "pion rocket" could help open up the Solar System with rapid and economical transport, *if* —a big if—the cost of generating antimatter in nuclear accelerators could be brought down substantially. Current accelerators generate antimatter by smashing high energy protons against metal targets. Antiprotons arise with much less than 0.5% efficiency—in effect, from raw electric power—but perhaps a way will be found to improve this figure. Forward suggests that $10 million per milligram is the crossover point for direct competition with current chemical propulsion systems. But its present cost is $100 billion per milligram, seemingly a long way to go!

As for interstellar rockets, the appeal of antimatter is obvious. Matter-antimatter annihilation is on the order of 1000 times more energetic than fission reactions and 100 times more energetic than fusion reactions, but all that glitters may not be gold.

Relativistic vs. Nonrelativistic Rocketry

For starship velocities that are tiny compared with the speed of light, the ordinary equations of Newtonian mechanics given earlier suffice to describe the performance of a rocket. But if a rocket's velocity becomes a significant fraction of light velocity, Einstein's equations of Special Relativity are required to describe its dynamics adequately. Remarkably, Einstein's theory of Special Relativity (which includes the famous $E = mc^2$) is derived from a *single* observational fact that has withstood every experimental test, namely: The speed of light in empty space has the same value for all observers, regardless of their speed toward or away from the light beam. (Actually, a second more fundamental postulate is also made in Special Relativity, namely, that "all inertial reference frames are equal with respect to the observed laws of physics." But the constancy of the speed of light is the one that is really dramatic.) Non-intuitive and contrary to "common sense," this is clearly not the case for rocks thrown at roadside signs from cars going at different speeds, but it is always true for photons of light. That is just the way the world is put together. Not surprising then, many strange results follow from this postulate of relativity (the constancy of the speed of light): phenomena like the mass and energy of an object increasing with velocity, the shrinkage of objects with increasing velocity, and the slowing down of "shipboard" time.

In the equations of relativity, the factor, γ (gamma), appears often:

$$\gamma = \frac{1}{\sqrt{1 - \frac{V^2}{c^2}}}$$

where V is the velocity of an object relative to some observer's reference frame and c is the velocity of light—about 3×10^8 m/sec. If V is very small compared with c, as it is for all velocities encountered in everyday life, then γ is close to 1.0. However, as V increases, this relativistic factor becomes significantly different from 1.0, and the equations of rocket performance must change correspondingly. To get an idea of the kinds of departures expected in different flight regimes, consult Table 3–2.

Gamma is important, for example, in describing how the *inertial mass* of a body increases with speed: simply multiply the *rest mass* of a body (denoted by m_o, the mass observed when $V = 0$) by gamma. So the inertial resistance to acceleration by an applied force increases dramatically with increasing velocity. At 0.99 c this is a seven-fold effect and at 0.9999 a 70-fold one, but as high as 0.5 c the effect is only 15% and at 0.1 c the effect on mass is less than 1%. Since 15% effects do not concern us much in reckoning the *feasibility* of starflight, a general rule is that up to 0.5 c, ignore the effects of relativity. On the other hand, to actually carry out an *accurately guided* interstellar mission, the effects of relativity would certainly have to be considered even at very low velocity.

Table 3-2
Relativistic
Flight Regimes

V/c	γ
0.0001	1.0000000
0.001	1.0000005
0.01	1.000050
0.02	1.000200
0.05	1.001252
0.1	1.005038
0.2	1.020621
0.3	1.048285
0.4	1.091089
0.5	1.154701
0.6	1.250000
0.7	1.400280
0.8	1.666667
0.9	2.294157
0.95	3.202563
0.98	5.025189
0.99	7.088812
0.999	22.366272
0.9999	70.712446

Table 3-3 Required Mass Ratios of Different Rockets for One-Way Proxima Centauri Fly-Through Mission (4.3 ly)

Specific Impulse (sec)	Overall Mass Ratio
Cruise Velocity = 0.05 c (Trip time > 86 years, including boost phase)	
500	1.3×10^{1328}
1,000	1.1×10^{664}
5,000	6.5×10^{132}
10,000	2.6×10^{66}
50,000	1.9×10^{13}
100,000	4.4×10^{6}
200,000	2.1×10^{3}
Cruise Velocity = 0.01 c (Trip time > 430 years, including boost phase)	
500	4.2×10^{265}
1,000	6.5×10^{132}
5,000	3.7×10^{26}
10,000	1.9×10^{13}
50,000	453
100,000	21.3
200,000	4.61
Cruise Velocity = 0.005 c (Trip time > 860 years, including boost phase)	
500	6.5×10^{132}
1,000	2.6×10^{66}
5,000	1.9×10^{13}
10,000	4.4×10^{6}
50,000	21.3
100,000	4.6
200,000	2.2

So how does relativity affect rocket performance? Many theorists have devoted volumes of ink to deriving the equations for relativistic rockets (44–54). Their work is historic and useful, but because relativistic dynamics is likely to play a fairly small role in starflight of the next 100 to 150 years, we summarize only a few important conclusions in Technical Note 3–3. These concern mass ratios, final velocities, and transit times. We have also provided in Appendix 5 an abbreviated discussion of the sometimes controversial "Twin Paradox," a thought experiment that demonstrates the reality of time dilation and its practical application to advanced astronautics.

Comparative Rocket Performance for Starflight

Rockets will probably never be applicable to better than Type-3 starflight, that is, for transit times to the nearby stars of 500 to 2000 years. The mass ratios required for such flights, even with superb engine performance, are simply too enormous to be reasonable. Some exceptions to this assessment are high-performance fusion rockets and nuclear pulse propulsion vehicles, which are certainly rockets of a kind (Chapter 4). To get an idea of these unreasonable, though in some (not all) cases *theoretically possible* initial/final mass ratios, examine Table 3–3. The table, computed from the nonrelativistic rocket equation (Technical Note 3–2), shows that even for a high-velocity, nondecelerated fly-through mission to the nearest known star, overall mass ratios for all but high specific impulse fusion vehicles (100,000 to 200,000 second I_{sp}) are absurd or literally *impossible* because they sometimes imply use of more mass than exists in the visible universe! (There are an estimated 10^{75} to 10^{100} atoms in the visible universe. The payload of a rocket might be only one atom if the mass ratio were in the range 10^{75} to 10^{100}, clearly an impossibility.) (Note that the number of stages does *not* affect the required *overall* mass ratio.) As Leik Myrabo and Dean Ing have written, "Where starships are concerned, a four-digit I_{sp} is about as much help as a one-digit IQ"(56).

Chemical rockets, "conventional" nuclear rockets, and low-performance electric rockets are out of the question even for flight times approaching a millennium. And remember that this table applies to only one-way fly-through missions. One-way decelerated missions or round trips would impose additional extraordinary multipliers to already ridiculous mass ratios. It seems that fusion and pulsed fusion rockets, maybe antimatter, or extremely advanced electric propulsion rockets may be the only kinds of rockets that could ever reach the stars in less than a millennium. But, Horatio, or Dr. Purcell, we have dreamt of much more than rocketry in our philosophy!

Relativistic Rocket Formulae

For a single stage rocket with constant exhaust velocity, V_e, the relativistic rocket equation is:

$$\frac{V}{c} = \frac{R^{\frac{2V_e}{c}} - 1}{R^{\frac{2V_e}{c}} + 1}$$

(1)

The final vehicle velocity as a fraction of the velocity of light is given as a function of rocket mass ratio and exhaust velocity (expressed as a fraction of the speed of light).

Rockets for which relativistic effects may be important are likely to have a significant fraction of their rest mass, ϵ, converted to exhaust energy and vehicle energy when they finally burnout. The expression of exhaust velocity as a function of ϵ is:

$$\frac{V_e}{c} = \sqrt{\epsilon(2 - \epsilon)}$$

(2)

Values of ϵ, the fractional conversion to energy of propellant only, for different advanced rocket systems are:

Energy Source	ϵ
Hydrogen/Oxygen	1.5×10^{-10}
Complete fission	7.9×10^{-4}
Fusion	4×10^{-3}
Complete matter annihilation	1.0

Note that for $\epsilon = 1.0$, that is, a pure gamma radiation exhaust, $V_e = c$ from equation (2), as expected.

So for a single-stage relativistic rocket, use of equations (1) and (2) gives:

$$\frac{V}{c} = \frac{R^{2\sqrt{\epsilon(2-\epsilon)}} - 1}{R^{2\sqrt{\epsilon(2-\epsilon)}} + 1}$$

(3)

For an n-stage rocket, each stage having the same exhaust velocity and with each stage proportioned to have the same mass ratio at burnout, the expression becomes:

$$\frac{V}{c} = \frac{R^{2n\sqrt{\epsilon(2-\epsilon)}} - 1}{R^{2n\sqrt{\epsilon(2-\epsilon)}} + 1}$$

(4)

So much for rocket mass ratios and final velocities.

The kinematical equations which relate the time and position coordinates in two inertial (unaccelerated) frames of reference are straightforward. For simplicity, consider only one-dimensional motion. If frame S′ is moving with respect to frame S with velocity, V, in the positive direction along the x-axis, then:

$$x' = \gamma(x - Vt)$$
$$x = \gamma(x' + Vt') \tag{5}$$

$$t' = \gamma\left(t - \frac{Vx}{c^2}\right)$$
$$t = \gamma\left(t' + \frac{Vx'}{c^2}\right) \tag{6}$$

where γ is the factor,

$$\gamma = \frac{1}{\sqrt{1 - \dfrac{V^2}{c^2}}} \tag{7}$$

For a rocket moving with *constant acceleration*, a, due to thrusting in its *proper frame*, the dynamical equations have been integrated by Sänger (55) and others to yield an expression for total elapsed proper time, $\Delta t'$, since the beginning of the journey: time as measured on the rocket:

$$\Delta t' = \frac{c}{a}\cosh^{-1}\left(1 + \frac{aS}{c^2}\right) \tag{8}$$

where S is the distance of travel and $\cosh^{-1}$ is the inverse hyperbolic cosine function. For $aS/c^2 >> 1$, equation (8) becomes:

$$\Delta t' = \frac{c}{a}\ln\left(\frac{2aS}{c^2}\right) \tag{9}$$

The vehicle's velocity after accelerating for $\Delta t'$ and reaching distance, S, is:

$$V = c\sqrt{1 - \frac{1}{\left(1 + \dfrac{aS}{c^2}\right)^2}} \tag{10}$$

4 Nuclear Pulse Propulsion

We have for the first time imagined a way to use the huge stockpiles of our bombs for better purpose than for murdering people. My purpose, and my belief, is that the bombs which killed and maimed at Hiroshima and Nagasaki shall one day open the skies to man. . . .

Freeman Dyson, *Mankind in the Universe,* 1970

The vast majority of scientists have consistently refused to become interested in the technical problems of propulsion, believing that this was a job for engineers.

Freeman Dyson, "Death of a Project," 1965

Stealing Promethean Fire

Chemical rockets are severely limited by energy; for all their storm and fury they are quite puny when compared with the needs of interstellar flight. Nuclear fission rockets—potentially much richer in energy—are, like their chemical rocket cousins, severely constrained in performance by temperature limits. The most advanced imaginable materials and cooling systems place an upper cap on fission nuclear rocket specific impulse. Ion engines, on the other hand, are power-limited rather than temperature limited. In order to achieve high specific impulse, an ion engine must rely on a source of electricity with extraordinary power generating capacity per unit mass. Fusion rockets may offer some relief from these conflicting problems, but it remains to be seen whether a sufficiently high fusion burn-up fraction will ever be achievable in a reasonably low-mass engine structure.

It almost seemed that nature had "rigged the deck" with these Scylla and Charybdis problems: too much temperature or too little energy or power. To save the day, enter *pulsed* nuclear propulsion, an extraordinary concept that developed from nuclear weapons research during World War II.

But before nuclear weapons there were chemical explosives. According to a detailed history of nuclear pulse propulsion by Anthony Martin and Alan Bond, senior members of the British Interplanetary Society, before the rocket becoming the preeminent form of space propulsion, a nineteenth-century engineer, Hermann Ganswindt, and physicist, R. B. Gostkowski, conceived vehicles that would propel themselves with a series of chemical explosions (1). The twentieth-century rebirth of this idea occurred during and after World War II, when the awesome fury of nuclear explosions began to spark the imaginations of space ship inventors.

So much greater is the energy release in nuclear fission or fusion explosions than in chemical detonations—factors of millions—that nuclear pulse propulsion must be a highly regarded prospect for starflight. The velocity of debris in a nuclear explosion may range from hundreds to 10,000 or more km/sec, implying potential I_{sp} in the range 10^4 to 10^6! The explosive release of nuclear energy is a method of circumventing the problem of the temperature-limited controlled fission rocket engine and the problematic, low-thrust continuous fusion rocket. But how to channel the enormous energy of nuclear detonations?

The Impossible Spaceship

It might have seemed to the most sensible of the Manhattan Project scientists that nothing could survive the hellish temperatures of tens of thousands of degrees unleashed in the fireball of an atomic bomb explosion, millions of degrees at the very center of the explosion. But as nuclear bomb research advanced, weaponeers learned that some materials could survive the nuclear inferno, and methods of directing or shaping the forces of nuclear explosions became an intriguing concept. Perhaps a blast could be deflected or reflected. After all, armor-piercing shaped-charge explosions were not new in conventional weaponry, and shaped chemical explosions served to compress and trigger fissionable uranium or plutonium in atomic bombs themselves.

Apparently, the first to speculate about nuclear pulse propulsion was mathematician Stanislaw Ulam in 1946. In 1947, he and colleague F. Reines prepared a report on the idea at Los Alamos Scientific Laboratory in New Mexico, where the Manhattan Project had been conducted. The general concept: Imagine, if you will, a flat steel plate coated on one side, perhaps with a thin layer of graphite. In space, a fission or fusion bomb exploded several tens of meters from the plate would not neces-

EXTERNAL EXPLOSION SYSTEM

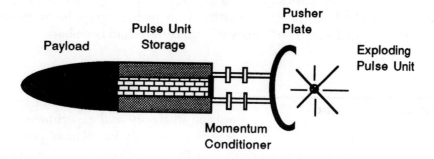

INTERNAL EXPLOSION SYSTEM

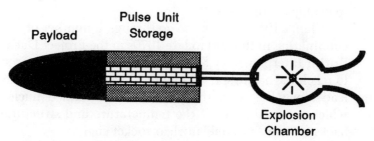

Figure 4.1 Nuclear pulse propulsion concepts.

sarily destroy it. Instead, vaporized bomb debris would impact and rebound from the plate, transferring momentum to it and causing it to move away from the site of the explosion. If the plate were attached to a spaceship by some kind of shock absorber, sequential explosions by rearward-ejected bomblets would propel the spacecraft (Figure 4.1).

In fact, at the Eniwetok Island nuclear test facility in the Pacific Ocean, an experiment conceived by physicist Lew Allen first offered direct proof of the merit of the nuclear pulse propulsion idea (2). The bomb experiment was code-named "Viper," with an explosive energy of 20 kilotons of TNT, roughly equivalent in yield to the bomb that destroyed Hiroshima. Two pumpkin-size steel spheres were covered with graphite and suspended from wires 10 m from the bomb. Though the wires were vaporized in the explosion, not so the steel spheres! Some time later and a good distance from "ground zero," the two steel spheres were found. The explosion blast wave had carried the spheres, much as an ocean wave supports a surfer, and their steel interiors were undamaged. Only a few thousandths of an inch of graphite were gone from their surfaces. The "impossible spaceship" had been born.

Early History of an Idea

The subsequent development of nuclear pulse propulsion in theory and experiment was so promising that it seems only by chance, politics, and extraordinary circumstances that *today* nuclear pulse powered spaceships are not zipping with ease across the Solar System. If nuclear pulse propulsion development had run its course, by now fast ships would be transporting people and instruments among the planets while tortoiselike chemical rockets stayed where they belong, in Earth orbit. By this time, much more serious thought would have been given to dispatching nuclear pulse spaceships to the stars.

Ulam's 1946 idea was to go through a number of incarnations before culminating in the mid to late 1970s in the Project Daedalus study by the British Interplanetary Society: the design of a robotic starship intended to reach Barnard's star in a mere 50 years. But there was one major false lead along the way: the internal or contained nuclear pulse system, which retained some of the temperature and structural limiting drawbacks of "conventional" nuclear rocket engines.

In the late 1950s and early 1960s, Dandridge M. Cole of the Martin Company and others at the Lawrence Livermore Laboratory examined the idea of detonating small nuclear bombs in the center of a spherical chamber to heat an inert propellant fluid injected into the vessel between explosions, diagrammed in Figure 4.1. The intensely heated propellant would expand through a nozzle to provide thrust. The Martin Company design required a 40 m diameter chamber, 0.01 to 0.1 kT (kiloton) nuclear charges exploded in it at 1 second intervals, and water as propellant, giving a specific impulse estimated to be little more than 1100 seconds. The similar Lawrence Livermore design, called Helios, envisioned hydrogen propellant injected into a 10 m diameter chamber in which 0.005 kT bombs were to detonate at intervals of 10 seconds. Nice try, impressive ships, but overall poor performance for the effort. You might say, they bombed!

External nuclear explosions proved to be far more rewarding. If a nuclear detonation occurs some distance from a vehicle, the temperature problems of internal explosions are considerably relaxed and much higher specific impulse is possible, perhaps 10^4 to 10^6 seconds! Temperature difficulties are reduced because of the short times—milliseconds—during which bomb debris interacts with the pusher plate. Experiments have shown that even ordinary metals, such as aluminum and steel, can withstand surface temperatures on the order of $80 \times 10^4 °K$ momentarily and sustain relatively minor ablation of their surfaces.

Cornelius Everett and Stanislaw Ulam at Los Alamos carried out the first analytic studies of the external nuclear pulse design and reported on their work in 1955. They designed a 12-ton craft with a 10 m pusher plate that carried from 40 to over 300 tons of bombs and inert propellant. It was to be equipped with 30 to 100 low-yield nuclear charges that would detonate 50 m from the craft at one second intervals and thus heat disk-shaped solid propellant masses, which were to be ejected in a coordinated fashion toward the bombs. The specific impulse of such a vehicle was thought to be in the range 1500 to 2000 seconds, still not much better than advanced "conventional" nuclear rockets. Nonetheless, on the basis of this work the then U.S. Atomic Energy Commission (AEC) applied for and was granted U.S. and British patents for the external pulse nuclear propulsion system.

The problem with external pulse designs, of course, is that much of the explosion's energy, in the form of bomb debris and heated inert propellant, is not intercepted by the pusher plate; a lot of the bomb's energy therefore goes to waste. And it is impossible with present techniques to design a pusher plate that will tolerate more than a few percent of the extremely high debris impingement velocity. T.W. Reynolds defined an "effective specific impulse," $(I_{sp})_{eff}$, for external nuclear pulse systems that depends on three factors: the "base specific impulse" (dependent on tolerable debris velocity at impact), the fraction of debris collimated to intercept the pusher plate, and the fraction of mass lost (3). Reynolds then published his probing insight into the nature of these factors. His findings are outlined in Technical Note 4–1.

Project Orion

The era of Project Orion (1958–1965) was truly the "golden age" of nuclear pulse propulsion. The project almost succeeded in igniting a new day in space propulsion within the Solar System, but in the end it was overwhelmed by a host of political problems. The $11 million effort by a team of some 40 technical participants carried out both experimental and theoretical studies. Orion's ascension and demise have been expertly chronicled by

---

Effective Specific Impulse for Nuclear Pulse

T.W. Reynolds (3) considered the system design of a generic external-pulse ablation propulsion system. His analysis considered a "base I_{sp}" that depends on the average propellant (debris) velocity tolerable by the pusher plate, a collimation effectiveness factor (f_c) of debris focused to intercept the pusher plate, and the fraction (f_m) of mass lost. He defined an effective specific impulse:

$$(I_{sp})_{eff} = f_c \, f_m \, (I_{sp})_{base}$$

The collimation effectiveness factor, f_c, has an upper limit of 0.5, rising with increasing pusher plate diameter and being higher for greater shaping or directivity of the nuclear charge. Reynolds made various gas dynamic and thermal transfer assumptions about the ablation process and showed that higher nuclear pulse energy yielded an f_m closer to 1.0 for a given pusher plate size. For a fixed pulse energy, he showed that increasing pusher plate diameter made f_m decline.

His primary conclusions: (1) For a given base specific impulse, pusher plate diameter and material characteristics, and inherent explosion shaping, there is an *optimum pulse energy* that gives maximum I_{sp} (2); increased pusher plate diameter gives higher I_{sp} for that optimum pulse energy.

---

one of its key participants, physicist Freeman Dyson (4). His "eulogy" concluded:

Orion had a unique ability to antagonize simultaneously the four most powerful sections of the Washington establishment. The remarkable thing is that, against such odds, with its future never assured for more than a few months at a time, the project survived as long as it did. It held together for 7 long years a band of talented and devoted men, and produced in that time a volume of scientific and engineering work which in breadth and thoroughness has rarely been equaled.

Orion began in 1958 as an effort by the General Atomics Division of the General Dynamics Corporation and was led by Theodore Taylor, a weapons designer who had worked at Los Alamos. During Orion's tortured existence, it was managed or comanaged by three federal agencies: DOD's Advanced Research Project's Agency (ARPA), the Air Force, and eventually, NASA. It was a case of too many cooks spoiling the broth and of the perennial conflict between civilian and military control and application.

Despite management problems, Orion succeeded in moving nuclear pulse propulsion from the abstract to the particular. Taylor dispensed with Everett and Ulam's separation of propellant and energy source, incorporating in the design "shaped" nuclear charges combined with plastic (perhaps polyethelene) inert material as "pulse units." This permitted, in theory, fully half of the pulse unit debris to be intercepted by the pusher plate.

Figure 4.2 is a schematic of the proposed Orion vehicle. Though many details of the Orion project remain in the murk of security classification, it seems that the team felt it had solved most of the key engineering problems for an external fission pulsed space vehicle. The team spent much effort in the design of a dual shock absorber system incorporating torus-shaped gas bags and pneumatic plungers. The researchers discovered that the pusher plate should be thicker in its center and taper to its edges. They also performed ablation testing (presumably not with live nuclear weapons, but with high-temperature plasma generators) to determine how the plate would fare.

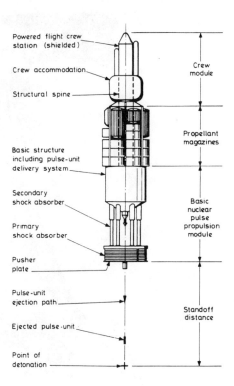

Figure 4.2 The Orion vehicle. (Courtesy JBIS)

Though no tangible nuclear spaceship emerged from Project Orion, a flight test vehicle, called "Put-Put," employing five chemical high-explosive shaped charges, was actually launched in southern California and rose to an altitude of 60 m (Figures 4.3 and 4.4), about as far *up* as Goddard's first liquid fuel rocket had traveled down range on March 16, 1926! But the project did end with a number of ship designs that were intended for various remarkably swift peopled missions to the planets, for example, to Mars and back in only 250 days! In 1968, before the first Apollo Moon landing, Freeman Dyson wrote, "We felt then that there was a reasonable chance that the U.S. could jump directly into nuclear propulsion and avoid building enormous chemical rockets like the Saturn V. Our plan was to send ships to Mars and Venus by 1968, at a cost that would have been only a fraction of what is now spent on the Apollo program" (5).

Orion project reports spoke of estimated specific impulse in the range 2000 to 6000 seconds with possible extension to the 10,000- to 20,000-

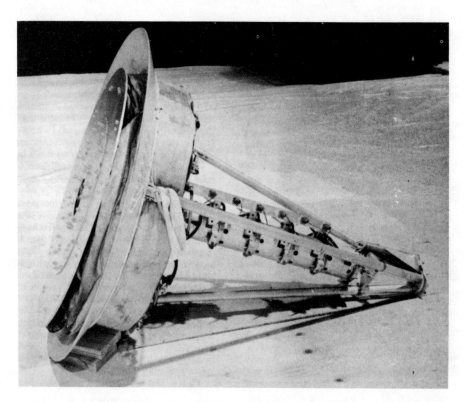

Figure 4.3   The Put-Put experimental nuclear pulse vehicle. (Courtesy JBIS)

Figure 4.4   Put-Put launch sequence. (Courtesy JBIS)

second range in the succeeding generation. A typical Orion vehicle design would use several thousand pulse units (0.01 to 10 kT) fired at repetition rates of 1 to 10 seconds, would accelerate at a significant fraction of 1 g, and would have a payload from hundreds to a thousand tons. Astronauts could carry everything with them on interplanetary missions, *including* perhaps literally many kitchen sinks. The project died because of the failure of nerve by political authorities, competition from chemical and conventional nuclear rockets, and the failure of the 1963 nuclear test ban treaty to unambiguously sanction nuclear detonations in space to propel peaceful space missions.

Dyson's Starships

Orion focused on nuclear fission pulse units only, but fusion explosions could not have been far from the designers' minds. In 1968, Freeman Dyson was first to propose publicly that nuclear pulse propulsion be used for starflight, and he suggested fusion pulse units (5). His conclusions were simple: The debris velocity of fusion explosions was probably in the range 3000 to 30,000 km/sec and the reflecting geometry of a hemispherical pusher plate would reduce that range four-fold to 750–15,000 km/sec (I_{sp} between 75,000 and 1.5×10^6). This made mission velocities of 10^3 to 10^4 km/sec possible.

To estimate the upper and lower limits of what could be done, Dyson considered two hypothetical kinds of fusion pulse starships. The more conservative design was *energy limited*, having a large enough pusher plate to safely absorb all the thermal energy of the impinging explosion, without melting! Dyson claimed the other design, maximum *momentum limited*, would define the upper region of performance. These were "thought experiment" monster starships (witness the summary Table 4-1) each designed to transport a colony of thousands of people to a

Table 4-1 Dyson's Starships

	Energy Limited Starship	Momentum Limited Starship
	Accelerate for 100 years at $3 \times 10^{-5} g$	*Accelerate for 10 days at 1 g*
Departure Mass:	4×10^7 tons	4×10^5 tons
Number of bombs:	3×10^7	3×10^5
Diameter:	20 km	100 meters
Velocity:	1000 km/sec	10,000 km/sec
Cost:	1 U.S. GNP (1968)	0.1 U.S. GNP (1968)

nearby star. It would take on the order of 1000 years for the energy-limited design to reach Alpha Centauri and a mere century for the momentum-limited vehicle.

Banning the Bomb: Fusion Microexplosions

A new era in thinking about nuclear pulse propulsion began in the late 1960s and early 1970s. Spurred by a new direction in research on controlled fusion for electric power generation, researchers literally "banned the bomb" from nuclear pulse propulsion. Instead of large, clumsy bombs, each of which would require fissionable material (whether in a pure fission bomb or in a fusion bomb trigger), they began to focus on igniting tiny "millikiloton" fusion microexplosions. They proposed to do this by focusing high-intensity laser light or relativistic electron beams onto small pellets of fusion fuel (6–11).

In theory, by lowering the energy of each fusion explosion the structural mass of a spacecraft could be reduced, provided that the laser or electron beam system mass were not prohibitive. Microexplosions also promised significantly reduced fuel costs because there would be no need for fissionable material or elaborate pulse unit structures.

First to write about fusion microexplosion propulsion was A.P. Fraas in the U.S. (1969–1971), who described a laser-ignited fusion pulse propulsion system, called "Blascon," with a remarkably low specific impulse (for any fusion system) of 3000 seconds (12). The pellets were of deuterium-tritium composition, a mere one centimeter in diameter.

The early work of Fraas and others was only the beginning. Soon microexplosion designs began to push toward theoretical specific impulse levels near 10^6—the "magic million," implying an exhaust velocity near 3% of light velocity! These proposals required that the old pusher plate become a powerful magnetic field, which would channel charged particles into an exhaust. And designers spoke of pulse repetition rates of hundreds per second. Converging laser beams would ignite the fusion pellets by "inertially" compressing and confining the fuel. Some of the energy of the microexplosions would be tapped electromagnetically to provide power for the lasers and the pusher plate magnetic fields, that is, a "bootstrap" process. These systems clearly have extraordinary design requirements and push technological limits, but that they could be put forth seriously at all is a tribute to the remarkable evolution in thinking about advanced propulsion. A vehicle propelled by a million-second I_{sp} engine could in theory visit any location within the Solar System in a matter of months!

To the Stars: Project Daedalus

Members of the British Interplanetary Society, who in earlier decades had been so prescient in describing what is now commonplace in astronautics, took up the challenge of fusion microexplosion propulsion and conducted the most elaborate study to date of a robot interstellar vehicle (13). From 1973 though 1978, the team of thirteen members worked on Project Daedalus, a two-stage fusion microexplosion spacecraft designed to send a scientific payload of 450 tons at 12% light velocity on a one-way, 50 year fly-through mission to Barnard's star's (presumed) planetary system, 5.9 light years distant.

The 10^6 second I_{sp} engines were to use pellets of deuterium and helium-3 fusion fuel; the latter component, because of its terrestrial scarcity, would have to be "mined" from Jupiter's atmosphere before the flight. Daedalus would accelerate for about four years under the incessant din of 50,000 tons of pellets ignited 250 times per second by relativistic electron beams. Total departure mass, fully fueled, 54,000 tons—almost all propellant (see Figure 4.5).

Epilogue

In ancient Greek mythology, Daedalus the clever inventor fashions wings and escapes from his island prison and flies to freedom across wine-dark seas. His son, Icarus, it must be said, crashed into the sea after disobeying his father's admonition about flying "too close to the Sun." In 1988, we have just witnessed the spectacular flight of a real Daedalus—a 35-kg human-powered aircraft designed and built by engineers and students at MIT. In a triumphant burst of human energy, the exquisitely designed propeller-driven craft was peddled by a single person, Kanellos Kanellopoulos, from the island of Crete to the island of Santorini, about 120 kilometers north across the Aegean Sea.

It took only 3500 years for the myth of Daedalus to be realized in plastic film, aluminum, and fibrous filament, not so long in the cosmic run of things. So it is not too hard to imagine, with the much greater acceleration of technology in modern times, that before many more centuries—perhaps only decades—another "real Daedalus" will set out, this time for the stars, seeking freedom from imprisonment in an island Solar System. There is at least a chance that the new Daedalus would be powered by fusion "fire," kindred to the power of stars, and no longer a threatening destroyer of humankind.

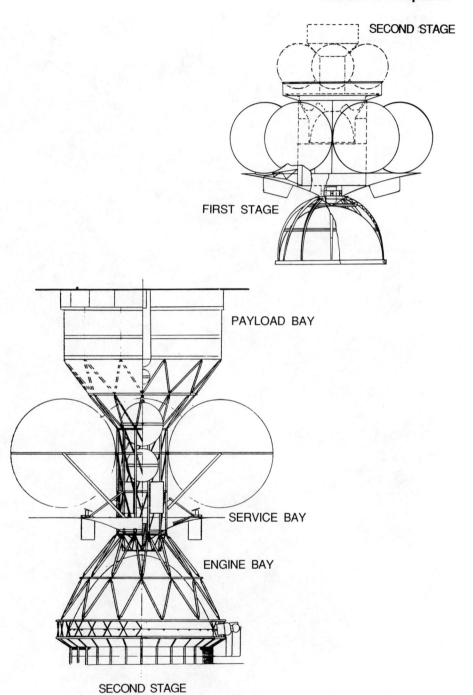

Figure 4.5   The Daedalus interstellar probe. (Courtesy JBIS)

SECOND STAGE

FIRST STAGE

PAYLOAD BAY

SERVICE BAY

ENGINE BAY

SECOND STAGE

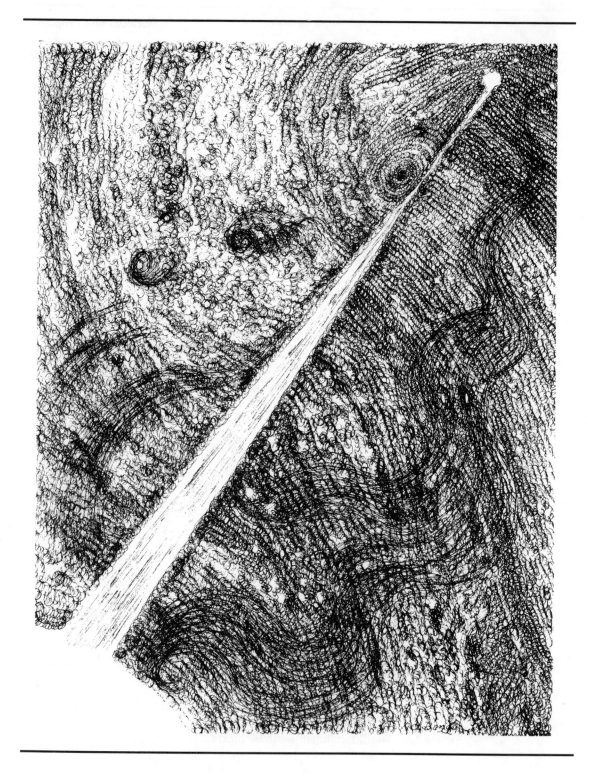

5 Beamed Energy Propulsion

I stared into the sky,
As wondering men have always done
Since beauty and the stars were one,
Though none so hard as I.

Ralph Hodgson, 1871–1962,
The Song of Honor

Among the stars lies the proper study of mankind; Pope's aphorism gave only part of the truth, for the proper study of mankind is not merely Man, but Intelligence.

Arthur C. Clarke, *The Exploration of Space,* 1951

Cutting Rocketry in Half

"Leave your energy source at home and leave the arriving to us." This is an appropriate slogan perhaps for propulsion systems that do just that: sever the energy producing apparatus from the starship. The result: much less mass to be accelerated. By beaming electromagnetic energy from the Solar System—light, microwaves, etc.—we could drive a ship to the stars by the pressure of radiation alone. But beamed power for starflight carries a potential *disadvantage:* the need to maintain economic and political stability on the homefront during a possibly very long acceleration period and perhaps during deceleration as well. But if our descendants were blessed with a civilization rich and stable enough to beam vast amounts of energy into the depths of space for several decades, it is possible that beamed power propulsion would make feasible that most difficult of interstellar voyages: the round-trip mission carrying people.

Beam-riding as a road to the stars appeared in a popular article before it emerged in the journals of science (1). Shortly after the invention of the laser in 1960 at Hughes Research Laboratories, physicist Robert L. Forward, then at Hughes, suggested using a space-based laser powered by the Sun to push a thin solar sail, a "light sail," in this case.

But he initially thought the concept would be too difficult to carry out in practice and so turned his attention to other interstellar propulsion ideas.

Another propulsion method that Forward considered at the time was *Lorentz force turning*, actually a method of modifying or curving an otherwise straight interstellar trajectory (3). Equipped with long wires that were charged to high voltages, an accelerated starship could use its electromagnetic interaction with the weak interstellar magnetic field to turn through a gigantic curved path. Forward suggested that the wires could be doped with suitable radioactive isotopes, which through their decay would charge the wires. By combining the two approaches, laser sailing and Lorentz force turning, it is possible to conceive of *two-way* interstellar transport with no fuel cost beyond what is required to achieve initial flight speed. In 1969, Canadian scientist Philip Norem was first to recognize this marvelous synthesis of two disparate concepts (4).

Interstellar "riders of the shining beam" actually fall into two basic categories that either: (1) Use the direct push of electromagnetic radiation; or (2) Use beamed power to energize and expand propellant that is carried along or obtained from the interstellar environment.

Beam-Propelled Light Sails

A two-way mission to Alpha Centauri, based on Forward's and Norem's laser sail ideas, would employ a solar-pumped laser orbiting within the inner Solar System, the closer to the Sun the better (Figure 5.1). Norem had optimistically assumed that 50% of the sunlight intercepted by the laser's solar energy collecting array could be converted into the laser beam's monochromatic, narrowly collimated radiation. This was perhaps overly optimistic, according to Forward's 1984 review of the efficiencies of existing lasers (5). The region of the electromagnetic spectrum that is visible to the unaided human eye extends from a wavelength of about 0.4 μm to 0.7 μm, where "μm" signifies micrometer or 10^{-6} meters. Forward claimed that carbon dioxide gas lasers, which generate 10.6 μm infrared light, have a 10 to 20% conversion efficiency; solar-pumped iodine lasers are 16% efficient and generate 1.315 μm light in the near-infrared spectrum; and free-electron lasers, which can operate at any selectable wavelength, have efficiencies of 30–50%.

Forward and Norem provided no engineering details of the optical system required to transmit light to a starship that would ultimately be light years away from the laser. However, they did apply the so-called *Rayleigh criterion* from basic optical diffraction theory (Technical Note

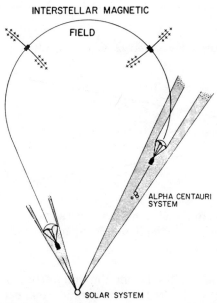

INTERSTELLAR MAGNETIC

FIELD

ALPHA CENTAURI
SYSTEM

SOLAR SYSTEM

Figure 5.1 Round-trip laser sail flight to Alpha Centauri. (Courtesy Robert L. Forward)

5–1), to determine the required diameter of the laser transmitter. For 0.5 μm yellow light, a 270-km diameter laser transmitter is needed to project a beam and have it just fill a light sail four light years removed. A 400-km aperture transmitter is required to beam to a ship near Tau Ceti, 11 light years from the Sun.

Norem's design for the inhabited starship was more specific than for his laser transmitter: a total ship mass of 3000 tons apportioned equally among payload, sail, and suspension wires, which would support the sail and connect it to the payload. He postulated a metal light sail of ultrathin titanium, a mere 0.2 μm thick, but 10 to 40 km in diameter. Norem optimistically suggested that this sail could operate continuously at a temperature of 1000°K, while still preserving a 99% reflectivity to the monochromatic laser light. The 1% of the light energy absorbed (not reflected) would be responsible for this extraordinary heating.

Using equations for the pressure of laser light photons impinging on the light sail (Technical Note 5–2) and assuming a glaring laser light intensity of 570 kilowatts/m² on the sail, Norem projected that his starship would accelerate at 0.4 m/sec² (about 0.05 g) and would approach relativistic velocities 27 years later. (By comparison, solar light intensity in space at Earth's distance from the Sun is roughly 1.4 kw/m².) A laser

Rayleigh's Criteria

The laser beam and sail geometry is schematically:

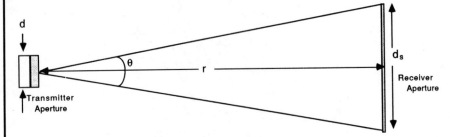

where θ = diffraction limited beam divergence angle
  r = separation between light source, S, and the sail
  d_s = sail diameter.

The beam of radiation just fills the sail, thus minimizing energy that bypasses the sail. For a diffraction-limited radiation source (monochromatic) and θ a small angle, Rayleigh's criterion that determines the required diameter of the radiation transmitter is:

$$\sin \theta \approx \theta = 1.22 \lambda/d$$

where λ is the radiation wavelength and d is the diameter of the optical system that collimates the beam. The larger the transmitter and the smaller the wavelength, the sharper the beam collimation. But from the geometry of the figure,

$$\sin \theta \approx \theta = \frac{d_s}{2r}$$

Therefore,

$$d_s/2r = 1.22 \, \lambda/d.$$

The distance at which the beam will just fill the sail is then:

$$r = d_s \, d/2.44 \, \lambda$$

If radiation is not to be wasted in "bypass" as the sail travels away from the radiation transmitter, either the sail must unfurl (expand) with time, the transmitter must grow in aperture, or both must be large enough at the outset to be good for all ranges during the mission.

The Pressure of Laser Photons

The energy of a photon of light is:

$$E = h\nu,$$

where h is Planck's constant and ν is the radiation frequency. The wavelength of a light photon and its momentum, p, are related by:

$$\lambda = h/p$$

But the speed of light, c, is related to ν and λ:

$$c = \nu \lambda$$

Combining these three equations, we find:

$$p = E/c$$

Thus, if a beam of total photon energy E_b is completely absorbed by a solar sail, the momentum lost by the light beam and gained by the sail is:

$$\Delta p_i = E_b/c$$

This momentum transfer corresponds to a totally inelastic collision by the light quanta. For totally elastic collisions of the photons (100% sail reflectivity), the photon paths are exactly reversed in direction at the sail and the sail gains momentum:

$$\Delta p_e = 2\, E_b/c$$

The starship's momentum change per unit time is:

$$\dot{p}_e = M_s \dot{V}_s$$

where M_s is starship mass and $\dot{V}_s$ is the starship acceleration. Substituting, we derive an expression for the acceleration

$$\dot{V}_s = \frac{2E_b}{M_s c}$$

generator system power of about 10^{16} watts would be required—more than 1000 times the present world energy usage, or a power equivalent to the explosion of one 100 megaton thermonuclear weapon every 2.8 minutes! If the laser array were powered by 10% efficient solar photovoltaic cells located near the planet Mercury, the diameter of the energy gathering array would be 1000 km. Without question, only an energy-wealthy interplanetary civilization capable of engineering planetary scale structures could deploy such a device.

Approaching the destination star system, the intrepid explorers would move their ship out of the beam, fold the light sail, and deploy 100 metallic cables—each of 50,000 km length. With an electric potential of 800,000 volts and a charge of 3.7×10^4 coulombs, the starship would use the Lorentz force (Technical Note 5–3) to turn through a long loop behind the target star. Like sailors of old, the crew would later reel in the charged cables, redeploy the light sail, and emerge once again into the laser beam.

Using the pressure of the laser beam, the ship would then decelerate into the target solar system from behind it, quite literally a backdoor approach. After years or decades of exploration and colonization, the temporarily mothballed starship would reenter the laser beam, accelerate away from the star until cruise velocity was reached, and use Lorentz force turning to redirect the trajectory toward the Sun. Approaching the Sun, the electrically charged cables would be wound up a final time, the

Technical
Note
5–3

Lorentz Force Turning

As described by Robert Forward (3), a charged object moving through a magnetic field experiences a force at right angles to its direction of motion and the magnetic field. The magnitude of the Lorentz force is:

$$|\overline{F}| = |Q\overline{V} \times \overline{B}| = QVB\sin\theta$$

where Q = charge on the object, $\overline{B}$ is the magnetic field vector, $\overline{V}$ is the velocity vector, and θ is the angle between $\overline{V}$ and $\overline{B}$. If the magnetic field is uniform and $\overline{V}$ is perpendicular to $\overline{B}$, θ is 90° and

$$|\overline{F}| = QVB$$

The object will move in a circular orbit perpendicular to the magnetic field vector. Since the object moves in a circular path, using Newton's Second Law we can set the Lorentz force equal to the the object's mass, m, times its acceleration, V^2/r, for an orbit of radius, r:

$$\frac{mV^2}{r} = QVB$$

Therefore, the instantaneous radius of the trajectory of a charged starship is:

$$r = \frac{mV}{QB}$$

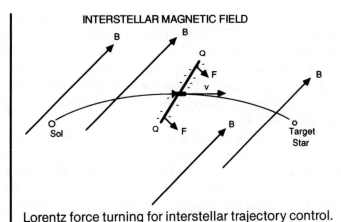

INTERSTELLAR MAGNETIC FIELD

Lorentz force turning for interstellar trajectory control.

sail redeployed, and the ship would reenter the beam to decelerate. Norem estimated that round-trip flight times of 75–150 years to nearby stars would be possible.

The most obvious drawback of the Forward-Norem concept is the need for a huge curving trajectory, light years in radius, to allow the ship to rendezvous with a tiny target in the depths of cubic light years of space: the 100 km diameter laser beam from the Solar System. A very small course deviation and the crew would be doomed. If the chance to decelerate were lost, the starship hulk might wander the galaxy forever. The depressing prospect of a galactic "Flying Dutchman" is further enhanced by uncertainties in our knowledge of the magnitude and variation of the interstellar magnetic field. These unknowns could well be fatal to the scheme.

Others soon considered modifications of the Forward-Norem proposal. Using X-ray lasers, as suggested by G. Marx and J. L. Redding (6,7), W. E. Moeckel considered 100-ton relativistic flyby probes, each requiring 10^{12} watts of beamed X-ray energy (8). In theory, the advantage of X-rays over visible light is the smaller transmitter aperture required for a particular narrow beam collimation. In practice, X-ray lasers may be much more difficult to implement even if they are smaller. J. H. Bloomer suggested using adaptive reflective mirrors near the Sun as a white-light alternative to the monochromatic laser (9). He proposed electrically charged or spinning membrane reflectors to control the shape or figure of the mirrors, thereby maintaining collimation. Freeman Dyson favored a system that would use beamed light sailing for acceleration and then interaction with the interstellar medium to decelerate a one-way interstellar colonization mission (10).

In 1984 Forward proposed an improvement on Norem's original idea that once again added to the credibility of round-trip interstellar travel (12). Forward considered an aluminum sail only 16 nanometers—billionths of a meter—thin, supported on a rotating wire truss that Eric Drexler had suggested. The truss would give such a thin sail its necessary rigidity (11). Forward recommended that the light sail temperature be kept below 600°K.

After demonstrating the superiority of a single large laser generator over a sparse coherent array of smaller lasers, Forward described the two-way mission illustrated in Figure 5.2. This approach would use a *Fresnel zone lens* or a *"Paralens"* (Technical Note 5–4) in orbit around the Sun to focus laser light on a multistage sail. With an initial sail diameter of 1000 km and a laser power of 4.3 X 10^16 watts, the ship would

---

Technical Note 5–4

The Fresnel Zone Lens

The "O'Meara Paralens" shown here was described by Forward (5). O'Meara, a colleague of Robert L. Forward, called the device a *paralens* because of its visual similarity to parachutes with annular holes that are used to decelerate some racing cars and high-performance aircraft upon landing.

Radial spokes and cross-members of wire define the shape of the lens. Attached to the delicate support structure are concentric rings of ultrathin plastic sheets. The sheets are of mathematically precisely tailored width for a given wavelength and they alternate with annular voids.

Forward analyzed the lens by deriving the relationship between its focal length, f, laser wavelength, λ, and the radius, r_n, of the nth zone. For 1.0 μm light and a 1000 km diameter lens, he found the total number of zones to be:

$$N = \frac{r^2}{f\lambda} = 111,410$$

To give some idea of the zones of the lens: They are of unequal radial thickness; the radius of the central zone is 1.5 kilometer; the spacing between the two outermost zones, 2.25 meters; the total mass of the lens, approximately 5×10^5 tons. Forward proposed light-levitating the lens between the orbits of Saturn and Uranus to focus light from lasers based near Mercury onto the distant sail. And he proposed introducing phase shifts in the light, analogous to the phenomenon of "Newton's Rings." These would be caused by laser light traversing alternate layers of plastic and empty space in the lens. He would prevent this effect by selecting the plastic thickness such that the excess optical path length is a half wavelength of laser light.

When Matloff considered transmitting intensified sunlight to a solar-sail starship via a Fresnel lens he found that chromatic aberrations inherent in the lens render it of marginal use (16). A reflecting optical system would perform far better.

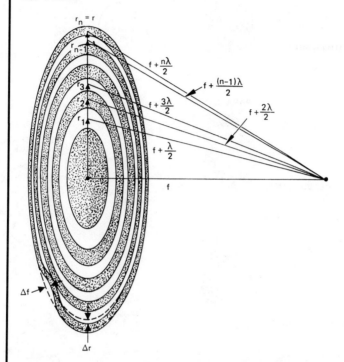

Fresnel zone lens or O'Meara paralens. (Courtesy Robert L. Forward)

approach Epsilon Eridani after a 23-year flight time (from the point of view of terrestrial clocks). Payload mass would be 3000 tons, including crew habitat, supplies, and exploration vehicles.

Nearing the destination, the sail would separate into an annular decelerator stage and a 320 km central "rendezvous" stage. Laser light would strike the decelerator and reflect off it onto the smaller sail still attached to the starship. After exploring and perhaps colonizing the Epsilon Eridani planetary system, the sail would again split up, this time into a 100 km return stage attached to the payload and a 320 km diameter annular accelerator stage. Laser light from the Solar System would reflect from the annular accelerator stage and push the disk-shaped return stage back toward the Solar System. Nearing home in the last phase of the mission, direct laser light would slow the return stage.

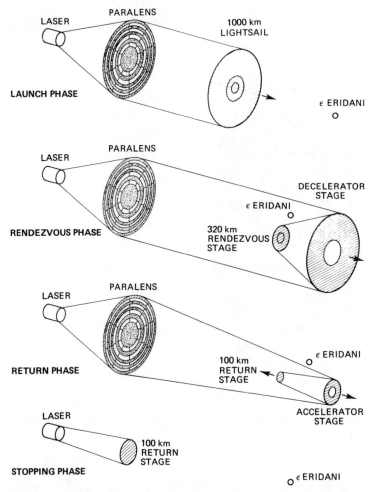

Figure 5.2 Segmented laser sail for round-trip interstellar flight. (Courtesy Robert L. Forward)

While carrying out a U.S. Air Force-funded study of futuristic space propulsion, Robert Forward developed yet another twist for beamed power systems. For part of this 1983 Air Force study, Forward traveled widely and interviewed many experts on advanced propulsion, among them Freeman Dyson of Princeton's Institute for Advanced Study. With a "back of the envelope" calculation, Dyson introduced Forward to the concept of the *perforated sail*. If that potent envelope still exists, we suggest that it be displayed in the National Air and Space Museum next to the Project Orion "Put-Put" prototype.

Dyson reasoned that a wire mesh would make a good reflector of

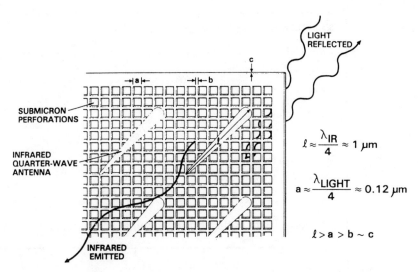

Figure 5.3 The perforated light sail. (Courtesy Robert L. Forward)

microwaves for a beamed power system if the size of the mesh openings were less than the incident wavelength. The same perforated reflector idea might be extended to sails for other parts of the electromagnetic spectrum. Perforated sails all would have the general benefit of significantly reduced mass—*areal density* or mass per unit area—from that of sails of unbroken material. It seems like cheating, but the laws of electromagnetic reflection do not lie; just ask any radio astronomer whose parabolic dish of "chicken wire" mesh efficiently reflects incoming waves. Forward's 1983 report illustrates a typical perforated light sail (Figure 5.3), and it also considers ways that such a "holely" light sail, the "holy grail of sails," could be fabricated.

Author Matloff has investigated interstellar flight applications of perforated sails designed not for single wavelength laser radiation, but for polychromatic or "white" light. He concluded that the assumption of equal transmission, reflection, and absorption is reasonable based on a comparison of experimental and theoretical results (13). Forward developed a monochromatic optical theory for perforated sails based on microwave measurements. It turns out that the monochromatic reflectivity of a perforated light sail can be high (Technical Note 5- 5).

Starwisp

Forward applied perforated sail ideas to the design of an extremely low-mass and therefore very fast, interstellar probe with the wonderful name "Star-

The Perforated Light Sail

Forward considered the reflectivity of a perforated light sail (14) and defined mesh parameter, k:

$$k = \frac{2h}{\lambda} \ln\left(\frac{h}{\pi b}\right)$$

where h = the maximum hole diameter
λ = wavelength of incident light
b = diameter wires that form the sail mesh

The reflectivity of the perforated sail as a function of the mesh parameter:

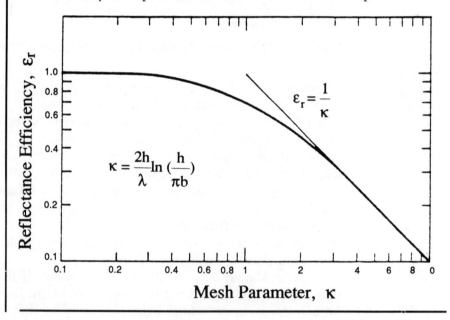

wisp," highly suggestive of the probe's insubstantial "spiderweb" appearance (14). He proposed borrowing the microwave beam of a *solar power satellite* (SPS) for a few days or weeks to drive the Starwisp reflector across the interstellar gulf. (SPS stations have been proposed for converting sunlight to electrical power in space, which would then be dispatched to Earth's surface via a tight microwave beam. Antenna fields on the ground would intercept the microwaves and transform their energy to electric power.) Forward proposed that the microwaves from the SPS first pass through an orbiting microwave Fresnel lens, which would focus the microwave beam on the receding probe—a mere spider-web of wires.

The Starwisp probe itself would be a mesh of ultrafine wires forming a disk one kilometer in diameter, yet its structural mass would be a mere 16 grams. Its payload: 4 grams of distributed and highly redundant "intelligent" microcircuitry coating the wires of the network, quite a challenge for the wizards of miniaturization! Starwisp would be akin to a flat neural net flying through space. Assuming an SPS beam power of 10^4 megawatts, Forward calculated that Starwisp would accelerate at 115g's and reach 20% of the speed of light within a few days. Upon its approach to Alpha Centauri about 20 years after launch, the SPS beam would be borrowed once again to flood the Alpha Centauri system with microwave power to energize the probe's data gathering and communications microcircuits.

Incorporating the payload of a solar-sail starship designed by the authors (15,16), Eric Jones of Los Alamos National Laboratory elaborated the Starwisp concept into a much more massive microwave-pushed interstellar ark (17). More recently, author Matloff has investigated the Fresnel lens for transmitting sunlight to an interstellar probe, furthering Bloomer's earlier ideas that relied on mirrors instead of lenses. James Early of the Lawrence Livermore National Laboratory considered combining laser light sails with other interstellar propulsion systems (18).

What would interstellar voyagers see of the laser beam that made their voyage possible as they looked back, perhaps wistfully, at the dimming Sun? If it had sufficient intensity, the monochromatic visible laser light from the Solar System would literally color the appearance of the star. This brings to mind the science-fiction novel, *The Mote in God's Eye*, in which Larry Niven and Jerry Pournelle describe the interaction of an alien race, the "Moties," and a human interstellar colony (19). A Motie sailing ship, powered by a visible-light laser transmitter, visits the human-colonized solar system. Because the Motie home star is visible as the revered "eye of God" in a prominent constellation in the sky of the human colony, the sudden appearance of a "mote" of laser light takes on deep theological significance.

Beamed Power Rockets and Ramjets

The other kind of beamed propulsion for starflight—the use of beamed power to energize rocket or ramjet propellant—seems much less elegant and fruitful. After all, a considerable benefit of beamed power is to reduce the mass that must be accelerated. To reintroduce the drawbacks of rocketry may be a disservice to the concept, yet a number of starflight researchers have been attracted to the idea. Laser-powered interstellar rockets grew out of the possible near-term application of lasers to boost payloads from the ground to low Earth orbit, a suggestion that Arthur Kantrowitz of the Avco Everett Research Laboratory first made (20). Leik Myrabo and Dean Ing in their imaginative review of the "future of flight" have discussed many aspects of this terrestrial and near-terrestrial application (21).

A. A. Jackson, IV and Daniel Whitmire considered an interstellar laser rocket, carrying out a relativistically correct analysis of Bloomer's earlier nonrelativistic treatment (22). They projected how an interstellar rocket carrying inert reaction mass would accelerate. The reaction mass —probably hydrogen—is energized by a laser based in the Solar System (Figure 5.4). Technical Note 5–6 presents a simplified analysis of the concept. Whitmire and Jackson have also considered relativistic interstellar travel using the laser ramjet, which gets its inert propellant or reaction mass from the interstellar medium(23)—shown schematically in Figure 5.5 and described briefly in Technical Note 5–7.

Author Matloff had once assumed that solar-electric powered ramjets would be useful only for auxiliary propulsion during the preperihelion phase of an interstellar solar sail mission (see Chapter 6 and Reference 24). But after the recent projection of electric propulsion performance by Graeme Aston at the Jet Propulsion Laboratory (25), we conclude that laser-electric propulsion might well have interstellar flight applications in the foreseeable future. Matloff's 1987 reevaluation of the laser-electric ramjet in light of Aston's projections (26) was optimistic about the drive's capabilities, work that has been expanded by the authors (27).

Consider a laser ramjet with a *ramscoop*, formed with a "modest" 1000-km radius magnetic field (see Chapter 8), and coursing through an interstellar medium with 10^5 ions per cubic meter. The starship's collector would receive laser energy and use it to increase the kinetic energy of

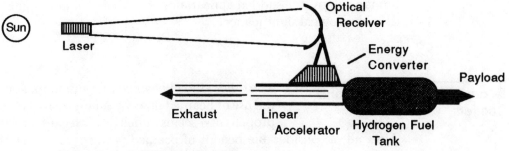

Figure 5.4 The interstellar laser rocket.

The Laser Powered Interstellar Rocket

The basic rocket equation is:

$$R = \frac{M_s + M_f}{M_s} = e^{\frac{\Delta V}{V_e}}$$

where R = the mass ratio
M_s = the empty ship mass
M_f = the total fuel mass
ΔV = the starship's total velocity increment
V_e = the exhaust velocity (as usual, in the reference frame of the ship)

If a laser with beam power, P, is aimed at the ship and converted to exhaust kinetic energy with efficiency, ϵ, we have the relation:

$$\frac{1}{2}\dot{m}V_e^2 = \epsilon P$$

where $\dot{m}$ is the rate of propellant flow into the exhaust. Therefore,

$$V_e = \sqrt{\frac{2\epsilon P}{\dot{m}}}$$

Combining this expression with the rocket equation above, the required mass ratio for a laser rocket to perform a mission with velocity increment, ΔV, is:

$$R = e^{\frac{\Delta V}{\sqrt{\frac{2\epsilon P}{\dot{m}}}}}$$

the ion exhaust. We assume that the laser light could be converted to exhaust kinetic energy with an efficiency of 50%. The laser power might be in the range 10^4 to 10^5 megawatts (of the same order as that proposed for Starwisp); the ion exhaust velocity, 5000 to 10,000 km/sec; and the total starship mass, 7.5×10^3 tons with a habitat payload of 3×10^3 tons. The required aperture of the Solar-System laser transmitter: 100 km for a 2 ly transmission distance and 1000 km for 20 ly. The laser ramjet may be of service for one-way colonization missions requiring a few centuries of transit time.

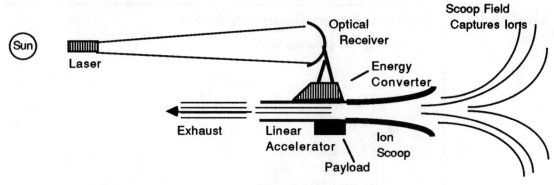

Figure 5.5 The interstellar laser ramjet.

A laser ramjet might use a solar sail (see Chapter 6) to boost it to an initial velocity of 0.003 C. Moving into the laser beam and starting the ramscoop and electric thruster, the ship would accelerate to 0.008 C in 85 years. After coasting toward Alpha Centauri for 400 years, the starship would decelerate using an electrostatic or electromagnetic drag screen (see Chapter 8). One-way trip time to Alpha Centauri: about five centuries.

Were it not for concerns about maintaining the homefront's support of the laser for half a millennium, it would be attractive to power the laser ramjet starship for the full duration of the mission, abolishing the "deadtime" of interstellar cruise. The colonists could then reach Alpha Centauri within about 350 years and Tau Ceti after 500 years. All stars within 21 ly of the Solar System would be accessible with trip times of no more than 750 years. It appears that the laser rocket would be superior to the laser ramjet for flights to Alpha Centauri or Barnard's Star, but for more distant destinations the ramjet is likely to be faster and therefore preferable.

The authors are currently investigating a laser-electric powered interstellar electric rocket fueled with hydrogen stolen from comets, one of the most abundant resources in the Solar System for that propellant. Another idea is to construct the laser energy collector of the rocket or ramjet—100 km in radius—of perforated sail material, optimized for efficient collection of monochromatic laser radiation. Although erosion of solar sails by interplanetary debris appears not to be a problem, sails deployed for decades in interstellar space on fast-moving vehicles may experience problems with erosion, but solutions may be possible (28).

Beam-riding technology might also nicely combine with interstellar solar sailing, the propulsion method closely related to beamed power discussed in the next chapter. Approaching the nearest point to the Sun

The Laser Ramjet

In our nonrelativistic analysis of the laser powered interstellar ramjet (27), we begin with basic momentum and energy conservation relations:

$$M_s \dot{V}_s = \dot{M}_f V_e \qquad \text{(momentum)}$$

and

$$\dot{M}_f V_s^2 + 2\epsilon P = \dot{M}_f (V_s + V_e)^2 \qquad \text{(energy)}$$

where M_s = ship mass

$\qquad \dot{M}_f$ = rate of propellant intake to the ramjet

$\qquad V_e$ = exhaust velocity (in this case, relative to the interstellar medium)

$\qquad V_s$ = ship velocity

$\qquad \dot{V}_s$ = ship acceleration

$\qquad \epsilon$ = laser/exhaust energy conversion efficiency

$\qquad P$ = laser power received by the starship

Rearranging and solving for the exhaust velocity:

$$V_e = -V_s + \sqrt{V_s^2 - 2\epsilon \frac{P}{\dot{m}_f}}$$

Substituting this result into the momentum equation, the starship velocity equation is:

$$V_s = \frac{\dot{M}_f}{M_s}\left(-V_s + \sqrt{V_s^2 + 2\frac{\epsilon P}{\dot{M}_f}} \right)$$

on its trajectory, its *perihelion*, a solar sail vehicle could detach an *optical transfer system*. Using techniques developed by Matloff and Ubell (29), the optical transfer system would decelerate and move into position to transmit more concentrated sunlight to the receding starship's solar sail. The optical transfer system would perhaps "light-levitate" in a stationary position relative to the Sun using techniques described by Forward (30). Beamed white-light solar sailing might therefore be competitive with some beamed laser power systems on one-way missions. The relatively short characteristic acceleration times of solar sails are certainly their significant attraction over laser beamed power. Problems of sail erosion and the difficulty of long-term support for huge laser power stations disappear with the solar sail.

6

Solar Sail Starships: Clipper Ships of the Galaxy

When will the third romantic age in the history of spaceflight begin? The third romantic age will see little model sailboats spreading their wings to the sun in space . . .

Freeman Dyson, *Disturbing the Universe,* 1979

Light boats sail swift, though greater hulks draw deep.
Troilus and Cressida, Act II, Scene 3

For I dipt into the future, far as human
eye could see,
Saw the Vision of the world, and all
the wonder that would be;
Saw the heavens fill with commerce,
argosies of magic sails, . . .

Ulysses, Alfred Lord Tennyson

Sunlight Surfing

"Ride sunbeams to the stars? Sail to Alpha Centauri? Be serious!" Or so would say the skeptic of using sunlight-pushed gossamer sails to reach the stars. But the idea of using light pressure to push thin, reflective films tethered to payloads by ultrastrong filaments is a venerable concept of space propulsion with roots in the nineteenth century theory of electromagnetism. The physicist James Clerk Maxwell in 1873 first showed that light rebounding from a mirror exerts pressure on it. Ethereal light *can* move matter, not with the fury of nuclear fire or chemical conflagration, but by gentle nudging in the vacuum of space. The sunlight pressure in the vicinity of Earth on a piece of sail the size of a football field is only about the weight of a small beach pebble. But such gentle force integrated over time can work wonders and generate an impressive final velocity.

In 1905 Albert Einstein contributed unknowingly to the future of solar sailing on the ocean of space by showing that quanta of light—photons—could possess momentum, as we have seen in the previous chapter. Simply divide the energy of a photon by the speed of light, c, to calculate the momentum it carries. If quadrillions of photons are impinging on a reflective surface, they are silently transferring momentum to it, just as though multitudes of small cannons were bombarding a wall with projectiles. A thin film of aluminized mylar, for example, can be pushed by sunlight like a wind-blown leaf. Its acceleration is not very great, but give it time, and like the tortoise it will win the race. The thinner the film and the less dense its material, the greater the acceleration it will achieve for a given flux of solar illumination.

In the early twentieth century, no one knew, of course, about aluminized mylar or the even 10 to 100 times thinner films that were to be conceived by space engineers. A few prescient Russians—astronautical pioneers—nonetheless dreamed of sunlight-driven spaceships long before a single chemical rocket had gone into space. Konstantin Tsiolkovskii and F. A. Tsander in the 1920s conceptualized the solar sail, observing that interplanetary voyaging might be possible with large reflective sheets of material pushed by sunlight (1,2). The idea of using solar sails to reach beyond the Solar System is of more recent vintage, but it rests firmly on the theory of solar sailing between the planets that a few physicists pioneered in the 1950s.

In the May 1951 issue of *Astounding Science Fiction*, Carl Wiley wrote what was possibly the first nonfanciful article about solar sails, "Clipper Ships of Space" (3). Post-Sputnik, in 1958, physicist Richard Garwin wrote the first serious technical article about solar sailing, concluding prophetically that "the method of propulsion is of negligible cost and is perhaps more powerful than many competing schemes" (4). Garwin cited examples of performance such as a satellite that could escape Earth's gravitational clutch within one week and a craft that could travel to Venus in less than a month, returning in only one week. In 1959 engineer T.C. Tsu at Westinghouse Research Laboratories made the first detailed studies of the spiraling interplanetary trajectories that solar sail ships would follow (5). Engineer-science fiction writer Arthur C. Clarke in a 1964 magazine story described a race to the Moon by "Sun yachts." One of the ships, *Diana*, went astray and inadvertently became the "first of all man's ships to set sail for the stars" (6).

The term "solar sail" may be a misnomer, for it could give the mistaken impression that the so-called "solar wind," made of charged massive particles that continuously emanate from the Sun, is the agency of push. The solar wind, in fact, contributes negligibly to the environmental pressure on a space sail. The term "light sail" would be more appropriate perhaps, but then the source of propulsive force might equally well be a bank of high-energy lasers blasting the sail as in the previous chapter's discussion of beamed power propulsion. We choose to maintain the historically used term, solar sail. This propulsion system pushes the design philosophy of decoupling the energy source from the propellant to an extreme. As with laser-driven light sails, there is no real propellant—just rebounding photons obtained free from nature. But with the solar sail, even the energy is free because it comes from the natural thermonuclear reactor—Sol—that dominates our environment. Since the "propellant" of a solar sail—sunlight—is free, efficiency is from one point of view 100%, though energy conversion from sunlight to spacecraft kinetic energy is much less than 100% efficient. (Technically, *specific impulse* is infinite for the solar sail. Specific impulse, again, is defined for rockets as the ratio of thrust to the propellant weight flow rate, the latter obviously being zero for a solar sail.)

When humanity gets its spacefaring act together, solar sails will have undoubted utility in solar system exploration and commerce, but will they, as Clarke suggested, be applicable to interstellar flight? As a sail is boosted outward, the solar light intensity declines as the *inverse square* of the distance from the Sun. The Sun grows ever dimmer. If the sail does not accelerate fast enough when it is near the Sun, it will miss the chance to gather the energy it needs to achieve a high terminal velocity enroute to the stars. Can a solar sail be made to accelerate quickly enough to make reasonably brief journeys to the stars? Probably yes, but the answer depends in part on the kinds of materials that may be developed for solar sails and how precariously close to the Sun we dare unfurl them. We have calculated that certain kinds of automated probe missions to Alpha Centauri could be as brief as several hundred years with "state of the art" sail materials, that is, speeds several percent of c might be achieved (7,8). (Note, however, that in the future the natural nearly linear motion of other stars may bring them closer to the Sun than Alpha Centauri is now, making transit then even quicker.) One-way missions bearing colonists to nearby stars would require a world ship—in effect, an interstellar ark in which the distant descendants of the space pioneers who set out initially would arrive at the destination (9). Solar sails are elegant, but they are definitely a subrelativistic solution to interstellar transport.

An interstellar ark mission using a solar sail would necessarily be lengthy because of the constraint of human acceleration tolerance and sail/tether stressing. A large ark, suspended from a sail by gleaming diamond filament cables could not reach Alpha Centauri in much less than 1000 years (10). But space habitats such as those proposed by Gerard O'Neill and other space colonization advocates may be sociologically reliable enough to endure a millennium journey.

Furthermore, if humanity eventually becomes a solar system-wide civilization that survives for some four billion years, it may use its otherwise malevolent Sun—then in its bloated red giant phase—to sail faster to distant stars. The intensity of sunlight follows an inverse-square law of decline outward from the Sun. In other words, as the distance between Sun and an observer is doubled, sunlight intensity decreases by a factor of four. If an interstellar solar sail is deployed 1.5 million km from the center of the present-day Sun (only about one solar radius from its surface), the sunlight intensity has fallen by a factor of four when the sail has been blown by light pressure to 3 million km from the center. For a red giant star, the solar sail "runway" will be much longer because of the greater size of the star, so red giants make ideal springboards for interstellar solar sailing. The solar sail may be just what a Kardashev Type-II civilization needs to escape the dire consequences of its expanding parent star. (A Kardashev Type-II civilization commands a substantial fraction of the energy output of its sun.)

The Basic Physics of Solar Sailing

Light pressure on a reflecting sail is of course a function of distance from the Sun, but it also depends on how reflective the sail is. A completely absorbing sail (a "black body" sail) with light incident perpendicular to its surface of intensity, I (watts/ sq. meter), experiences a pressure, p_s, of magnitude:

$$p_s = \frac{I}{c}$$

where c is the velocity of light.
If the sail is *100% reflective*, then:

$$p_s = \frac{2I}{c}$$

Near the Earth, I is about 1400 watts/m². The solar radiation pressure on a thin piece of aluminized mylar could produce a force several times the gravitational attraction of the Sun on that piece of plastic.

For sail reflectivity other than 100%, k between 0 and 1.0, the formula for pressure on the sail material is:

$$p_s = (1 + k)\frac{I}{c}$$

So if a sail is 95% reflective, the solar radiation pressure will be reduced by about 2.5%. Because the Sun's light decreases in intensity as it spreads outward, it will be reduced according to the inverse square of the distance, r, from the Sun —that is, by $1/r^2$. So the pressure varies with distance as:

$$p_s = \frac{(1 + k)I_o}{r^2 \quad c}$$

where I_o is the solar light intensity at the surface of the sun. (The formula is reasonably accurate only for r >> Sun's radius.) A few more of the basic formulae governing solar sails are presented in Technical Note 6–1.

Technical Note 6–1

Sunlight Pressure: Basic Theory

Elementary units of light, photons, carry momentum as well as energy, as we saw in Chapter 5 (see Technical Note 5–2). The momentum, p, of a solar photon is equivalent to its energy divided by the speed of light: $p = E/c$. At the dawn of the space age, T.C. Tsu analyzed interplanetary solar sailing, using the following relationship for light pressure on a 100% reflective sail:

$$p_r = \frac{2S_r}{c}$$

where S_r is the *solar irradiance*. Applying the inverse-square law, we demonstrated (7):

$$S_r = (3.04 \times 10^{25})/r^2 \text{ watt/m}^2$$

where r is the separation (meters) between the sail and the Sun's center.

For a sail that does not transmit any sunlight though which has a reflectivity, k, we showed:

$$p_r = \frac{(1 + k)S_r}{c}$$

If the combined sail, cable, and payload mass is M and the circular sail's radius is R_s, the acceleration of the sail outward from the Sun is:

$$a_s = \frac{(1 + k)(6.3 \times 10^{17})R_s^2}{2Mr^2} \text{ m/sec}^2$$

The final interstellar departure velocity, V_∞, in meters/sec of a solar sail starting with velocity, V_o, at closest approach to the Sun we have calculated to be:

$$V_\infty = \left[[0.5(1 + k)(1.26 \times 10^{18} R_s^2 - 2.66 \times 10^{28}M]/Mr_o + V_o^2\right]^{1/2}$$

where r_o is the closest approach distance from the center of the Sun.

One of the authors has demonstrated another useful formula (10). For partially transmitting sails, k in the above equations is:

$$k = \frac{R_a}{(R_a + A)}$$

where R_a and A are the fractional sail reflectivity and absorption, respectively.

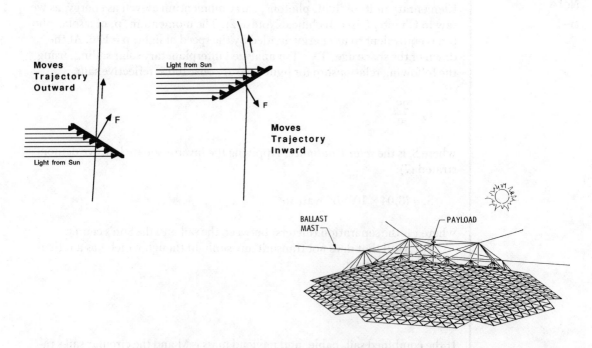

Figure 6.1 Basic solar sailing and a proposed interplanetary solar sail spacecraft. (Courtesy Robert L. Forward/Hughes)

It may seem impossible that a sail can propel a craft from an outer to an inner planet because the direction of solar photons is of course outward from the Sun. However, by simply tilting the sail surface appropriately it is possible to produce a component of force that is either in the direction of orbital motion or against it (see Figure 6.1). By *decreasing* its orbital velocity, a solar sail spacecraft can spiral in toward the Sun— almost like a sailboat tacking into the wind (but with no keel!). Another interesting feature of solar sailing is, with the right adjustment of sail area, the ability to exactly counterbalance solar gravity and to depart tangentially from circular orbit (see Technical Note 6–2).

Technical
Note
6–2

Sunlight Pressure and Newton's First Law

The effectiveness of sunlight pressure in propelling a solar sail radially outward from the Sun is apparent from an example. Consider a solar sail starship in circular orbit, r meters from the Sun's center. Before the sail is deployed, the solar gravitational force on the spacecraft, F_{grav}, and the centripetal force, F_{cent}, are equal. From the definition of F_{grav} and F_{cent}:

$$\frac{GM_sM_c}{r^2} = \frac{M_cV_{circ}^2}{r}$$

where G is the gravitational constant, M_s and M_c are respectively the Sun and the spacecraft mass, and V_{circ} is the circular orbital velocity of the spacecraft around the Sun. Rearranging this equation:

$$V_{circ} = \sqrt{\frac{GM_s}{r}}$$

Suppose that the spacecraft now deploys a solar sail such that the radial force, F_{rad}, is equal in magnitude to the gravitational force but in the opposite direction. The sail will leave the solar system with velocity, V_{circ}. This is because there is zero net force on the sail and the ship is therefore in dynamic equilibrium. Moreover, this condition continues because the solar flux and force of gravity decrease proportionally (as $1/r^2$) as the spaceship moves further away from the Sun. From Newton's 1st Law, the spacecraft will continue in a straight line at constant velocity.

For starflight, we generally want the full strength of the solar pressure force to be used, so the sail axis should point directly at the Sun. But as shown in Technical Note 6–3, there are some energy advantages to a tangential velocity boost near the Sun at perihelion.

Tangential Acceleration and the Powered Perihelion Maneuver

As well as imparting radial acceleration, a solar sail can accelerate tangentially. During a close solar flyby, taking advantage of a tangential velocity boost is more efficient than if performed in "free space"—that is, far from a gravitating body. Consider the situation depicted in the figure. Points 1 and 3 are far removed from perihelion, Point 2, that is distance R_p from the center of the Sun. The energy of the spacecraft at each of the three points is:

$$E_1 = \frac{M_c V_i^2}{2}$$

$$E_2 = \frac{M_c (V_p + \Delta V)^2}{2} - \frac{GM_c M_s}{R_p}$$

$$E_3 = \frac{M_c V_\infty^2}{2}$$

where V_i is the initial spacecraft velocity at Point 1, V_p is unaccelerated perihelion velocity, ΔV is the incremental velocity imparted by a tangential solar boost at perihelion, and V_∞ is velocity at Point 3 far from perihelion.

First examine the case of a perihelion pass with no propulsive maneuver ($\Delta V = 0$). $E_1 = E_2$ by the conservation of energy in a central force field, and

$$V_p^2 = V_i^2 + \frac{2GM_s}{R_p}$$

The escape velocity from the Sun, V_e, at perihelion distance, R_p, is $(2GM_s/r_p)^{1/2}$. Therefore,

$$V_p = \sqrt{V_i^2 + V_e^2}$$

Substituting this result into the first equation above,

$$V_\infty = [(V_i^2 + V_e^2)^{1/2} + \Delta V)^2 - V_e^2]^{1/2},$$

which reduces to

$$V_\infty = [V_i^2 + \Delta V^2 + 2\Delta V (V_i^2 + V_e^2)^{1/2}]^{1/2}$$

If ΔV were applied far from the Sun,

$$V'_\infty = V_i + \Delta V$$

For the solar boost close to the Sun to be more efficient than a powered maneuver done in "free space," $V_\infty > V'_\infty$. This is true as demonstrated below:

$$V_i^2 + \Delta V^2 + 2\Delta V(V_i^2 + V_e^2)^{1/2} > V_i^2 + \Delta V^2 + 2\Delta V V_i \qquad \text{and}$$

$$2\Delta V(V_i^2 + V_e^2)^{1/2} > 2\Delta V V_i$$

Because $(V_i^2 + V_e^2)^{1/2} > V_i^2$ in all cases, the powered tangential perihelion maneuver is more efficient than a corresponding free space powered maneuver.

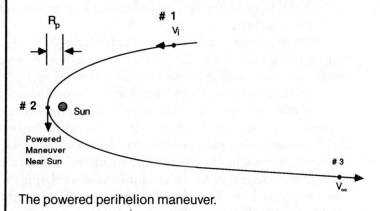

The powered perihelion maneuver.

Sail Materials

Humanity launched its first solar sail on August 12, 1960, the aluminized mylar balloon called the Echo-1 communication satellite. This 30m-diameter space balloon the U. S. launched was not intended to do any sailing, but sunlight pressure on its reflective, low-density hulk caused its perigee or closest approach to Earth to vary by as much as 500 km.

Aluminized mylar is, in fact, a good "low-tech" solar sail material, but it would not be suitable for interstellar missions. Its frontal area density—*areal density*—is simply not low enough for starflight but is suitable for interplanetary missions. Mylar film 0.0002 cm thick would have an area density of only about 5×10^{-3} kg/m² (kilograms per square meter). The thrust-to-weight ratio of a 100% reflective sail would only be about 0.0002 at Earth's orbit, but since the sail would be deployed in the weightless free-fall of orbit it would slowly accelerate. And lest this number seem depressingly low, remember that it is solar gravity, *not* Earth gravity that a solar sail opposes as it travels outward from the Sun.

The sunlight pressure force can be several to tens of times greater than the solar gravitational force on a thin sail.

A good figure-of-merit for a solar sail is its "lightness number" (L_n) or ratio of its acceleration by sunlight pressure to the Sun's gravitational acceleration. Since both solar gravity and sunlight intensity decline with an identical inverse-square relationship as a sail moves outward from or inward to the sun, this figure-of-merit remains the same for a given sail material. (The lightness number is not strictly an inherent property of the sail material but of the sail material/Sun combination. Another star with a different luminosity would give a particular material a different lightness number.)

We can imagine sail materials with better characteristics than mylar, sails with area densities 10 to 100 times lighter. But in order to deploy such sails, they will have to be manufactured in the vacuum of space because they are simply too fragile to be supported on the ground, to be folded into a payload, or to survive launch. Terrestrial experiments reported by Eric Drexler, formerly at the Space Systems Laboratory at MIT, created small pieces of ultrathin metallic sail using the technique of *vacuum deposition* (11). Investigators placed a glass slide coated with a thin film of detergent into a vacuum chamber. Using an electrical heater they vaporized small metallic samples, thus condensing a thin film of metal on the glass slide. Subsequent immersion in water of the metal/detergent coated slide caused the metal film to detach, leading to a metallic foil only a few hundred atoms thick, hundreds of times thinner than ordinary kitchen foil.

Space-manufactured solar sails using this deposition technique would have the advantage of the free vacuum environment of space. The sails would be 20 to 80 times lighter than mylar film sails, increasing maximum sail acceleration by approximately the same factor. Drexler has pointed out that one U.S. space shuttle payload bay could dispense enough materials to deploy one-hundred square km of solar sail reflector, though one or two additional shuttle launches would be required for manufacturing equipment.

Metallic sails can operate at temperatures in the range 1000–2000°K, thus making them ideal candidates for the close swing-by of the Sun required for an interstellar mission. A boron sail, for example, has a melting point of 2600°K and a density only 2.5 times that of water. (Boron is actually a "metalloid" with some semiconducting properties.) A boron sail 10^{-8} m thick (about 100 atoms across) would have only about two times the area density of the thinnest conceivable *solid* sail material (12). (Perforated sails—see Chapter 5—may have some utility in reducing mass density.)

Sail Design for Interstellar Missions

An interstellar solar sail will have to approach very close to the Sun to take advantage of the intense sunlight on a brief close encounter. To imagine the scale of such a grazing pass, first realize that the average distance from Earth to the Sun, about 150 million km or 1.0 astronomical unit (AU), is about 200 times the Sun's radius (about 0.7 million km or about 0.005 astronomical units). For tolerable interstellar transit times, a solar sail will have to approach within 0.01 AU of the Sun's center or within about one solar radius of the sizzling 6,000°K solar surface. According to Kraft Ehricke, 0.01 AU may be the closest feasible approach to the Sun without getting into extraordinary thermal shielding problems (13).

To get so close to the Sun requires, paradoxically, killing a significant amount of forward-directed orbital energy and then falling in (on a *parabolic* trajectory) or boosting in on a high velocity *hyperbolic* path. This cannot be done by solar sailing toward the Sun because far from the Sun solar intensity is not strong enough for the necessary deceleration/acceleration. Instead, the sail must be kept furled, boosted in compact form toward the Sun, and deployed unfurled near closest approach—*perihelion.* We have considered a variety of conventional and advanced propulsion systems to accomplish this, all of which are greatly assisted by boosting outward to one of the giant planets and then executing a propulsive maneuver during a high-velocity *gravity assist* swing-by of that planet. If a large velocity increment maneuver is performed with an advanced propulsion system near the giant planet, the resulting hyperbolic velocity relative to the Sun at perihelion will be significant (see Chapter 9).

The hostile thermal regime 0.01 AU from the Sun's center presents an extraordinary design challenge. There a flat sail pointed toward the Sun will receive 10^7 watts/m^2 of light energy, about 10,000 times the flux near Earth. Since both sides of a flat sail can radiate, a fully absorbing (*black body*) sail would get very hot—3100°K. If the sail were 90% reflective, then the temperature would be lower but a still elevated 1750°K. Extremely shiny sails with 95% and 99% reflectivity will have perihelion temperatures of 1500°K and 1000°K respectively, and modern optical coating engineering may well make sails with this high reflectivity feasible.

To maintain a thin, round sail in flat configuration, the sail would most likely have to be spun along its central axis so that the centrifugal force of rotation would maintain a flat sail profile. Forward has suggested that a robotic payload consisting of an intelligent array of extremely thin film microcircuits—akin to Starwisp's payload (Chapter 5)—could be deposited on the back of a reflective sail (14). This would

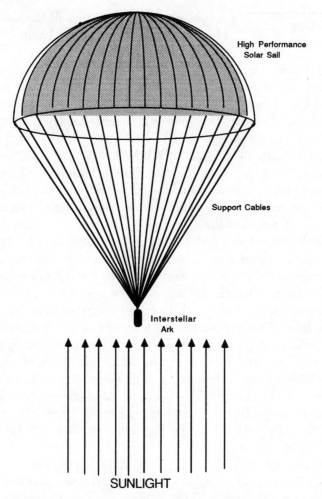

**High Performance
Solar Sail**

Support Cables

**Interstellar
Ark**

SUNLIGHT

Figure 6.2   Solar sail-towed interstellar ark.

permit an extremely low-mass flat sail configuration. However, for more conventional, concentrated-mass scientific payloads or human "ark" habitats, the sail configuration would have to be parachutelike and curved (see Figure 6.2). A spiderweb network of extremely high tensile strength filaments would form the backing of a curved sail and the filaments would extend from the sail perimeter to form tethers to the payload in tow. To keep such a sail open (unfurled) it would have to be rotated to balance the inward radial component of tension in the filaments. In fact, a substantial amount of energy would effectively be stored in the necessary rotation of a large curved sail. By despinning the sail during interstellar transit, colonists or automated instruments could tap a large reserve of energy for their needs, an idea that was suggested by the authors. Inward tilting sails are an alternative to spinning sails (7).

Diamond fiber filaments would be perhaps the strongest imaginable sail support network, though their material density would also increase the frontal mass loading—effective areal density—of a sail. It is possible to imagine a fiber network that segments the sail into small enough rectangular sectors such that the internal stress on individual panels of sail material would not tear a plastic sail, much less the stronger space manufactured metallic sails discussed earlier. Stress problems with a fine filament backing do not seem insurmountable, although they degrade optimum performance by increasing the sail mass.

One advantage of a curved sail is that it will have a lower temperature at perihelion. If at perihelion a 95%-reflective sail is aligned so that only 25% of the maximum possible solar flux strikes it, the perihelion temperature can be held to less than $1100°K–800°K$ for a 99% reflective sail. Fortunately, the duration of the high-temperature perihelion pass will be brief—typically less than one-half hour—as the sail approaches the Sun at high velocity and then accelerates to a distance millions of kilometers from the Sun. But the rapidity of the perihelion pass is both an advantage and a problem. In most cases the duration and intensity of the high temperature phase will be similar to reentering the Earth's atmosphere from low orbit, so an *ablative shield* may be adequate to thermally protect the payload. But there is no obvious way to deploy a thin sail 2 to 400 km in diameter in a matter of minutes or even seconds.

One solution is to provide the spacecraft with a "launching pad," an inert chunk of material, such as asteroidal rock, more massive than the sail and its payload and with dimensions similar to those of the sail. The asteroidal piece would serve as an *occulter*, behind which a fully or partially deployed sail could emerge quickly and gracefully. We have researched the complex logistics of such occulter maneuvers and find that they are feasible (7,8).

After the separation from the occulter, the solar sail starship begins to depart the Solar System with an acceleration, depending largely on the area density of the sail, from 1 g up to 700 g's. (Although human beings have withstood as much as 45 g's for fractions of a second, the record for extended high acceleration without ill effects or loss of consciousness is 17 g's for four minutes. Our calculations show that a human-occupied interstellar sail habitat would thus be limited to a final cruise velocity of 0.003 c.) Solar flux begins to decline and the area of the sail perpendicular to the Sun's direction can be increased to maintain adequate acceleration. Though this requires changes in sail radius of tens of kilometers per hour, the operation of winding and unwinding support cables seems feasible.

Icarus-1: A Solar Sail Interstellar Probe

In principle, twentieth-century technology would be capable of launching an automated scientific probe to Alpha Centauri that would take only about 350 years to get there. Enroute the probe could radio back valuable scientific information about the interstellar medium. We realize that no nation presently has the motivation to engineer such a mission, but that *Icarus-1* is even theoretically possible is an exciting prospect.

A sail/payload with a frontal density loading of 2×10^{-5} kg/m² and a sail radius of one km would have a mass of only 63 kg. If the probe were directed from the vicinity of Earth into a near parabolic orbit of the Sun and approached the Sun closely within one solar radius, it could unfurl behind an occulter, experience a maximum acceleration of 440 g's, and achieve a final velocity of about 0.012 c. Flight time to Alpha Centauri would be about 350 years. And let us hope that *Icarus-1* in nearing the Sun before being bound for the stars would suffer a better fate than Icarus, the son of Daedalus, who according to ancient Greek myth flew too close to the Sun and perished in the sea.

During interstellar cruise, the 10 kg or so of payload could monitor the interstellar medium, using microcircuit advanced sensors. Thin-film electro-optical components could be deposited directly on a sail to act as a large observing array at destination. V.R. Eshleman has suggested that an interstellar probe departing the Solar System could search for signals from extraterrestrial civilizations using the Sun's gravitational field to amplify radio emissions for stars that the Sun occulted (15).

Sark-1: A Solar Sail Interstellar Ark

Robot starprobes may unfurl their delicate wings to soar on the Sun's breath, but will people ever attempt the same ride? If they ever do, the first ones are likely to be colonists on a one-way mission. The requirements for a sail interstellar ark suitable to carry a "modest" crew of 1000 individuals to the Alpha Centauri system has been analyzed. The ark, affectionately called "Sark-1" (To recall the good ship Cutty Sark and the bottle of spirits it now adorns—lots of spirits will be needed on a millennium journey!) would be inhabited by 1000 persons each generously allotted 40 square yards of living space. Its basic toroidal cabin structure would weigh 2500 metric tons on Earth's surface. It would also be equipped with 1000 tons of atmosphere, plus additional supplies and power plant amounting to 2000 tons—a basic 5500-ton space colony, minus sail. But since additional supplies will be needed for setting up a space colony orbiting Alpha Centauri A or B, this space can be generously rounded to 10,000 tons.

For structural and safety reasons, the colony will be divided into six starships, each towed by a circular sail 380 km in diameter. We have optimistically assumed that microfine support filaments (cables) made of

Using a Solar Sail to Decelerate

The next chapter discusses interstellar *ramscoops* used to decelerate at a destination star. Unfortunately, the efficiency of a ramscoop falls rapidly as a starship's velocity decreases. Alternately, a solar sail could be unfurled as the destination star is approached and it could act as a deceleration device. Matloff and Ubell determined that the velocity of the sail decelerated starship a distance r_f from the center of the target star is given by (18):

$$V_f = \sqrt{V_c^2 - \frac{2(J - GM_{star}M_c)}{r_f M_c}} \tag{1}$$

where V_c is the interstellar cruise velocity of the starship, M_{star} is the star's mass, and

$$J = \frac{(1+k)}{2}(6.3 \times 10^{17})R_s^2 L_f \qquad (R_s \text{ in meters}) \tag{2}$$

In Equation 2, L_f is the ratio of the star's to the Sun's luminosity. Accelerations must be constrained to insure that the cables do not snap. One example considered by Matloff and Ubell, a ship decelerating from an initial velocity of 0.002c and a distance of 35 astronomical units (AU) from Alpha Centauri (virtually a twin of the Sun) comes to rest 0.7 AU from that star (18). The maximum deceleration on this 5×10^7 ton worldship during sail deceleration is 3.15×10^{-3} g's, reached when the starship is 7 AU from Alpha Centauri. As the ship approaches the star closer than 7 AU the sail is gradually furled to maintain constant deceleration.

diamond will be available when the mission departs. (Copper filaments would do, but would be less efficient and much less dramatic.) The sail area density (including cables) would be a gossamer-thin 4×10^{-5} kg/m². Maximum acceleration on the colonists as they boosted out from the Sun would be 14.6 g's from an initial perihelion velocity of 0.0014 c. They would take about 1350 years to reach Alpha Centauri if the mission began from a parabolic approach to the Sun—1200 years with auxiliary propulsion during the preperihelion phase.

A few tenths of a light year from destination, the sail is redeployed, electrically charged, and used to decelerate the starship electrostatically (see Chapter 8) to 0.001 c. For the final deceleration, the solar sail would be partially unfurled to use Alpha Centauri A or B's light for decelerating and maneuvering within the supposed planetary system (Technical Note 6–4).

Effects of the Interplanetary and Interstellar Medium

Once a sail has boosted its payload onto an interstellar trajectory, its propulsive function is typically finished—except for later electrostatic deceleration—and it may then be folded and stowed in more compact form during interstellar cruise, thus assuring that the interstellar medium will not degrade it. However, near the Sun during the propulsive phase there may be some concern about impinging space debris and intense solar magnetic fields. Although micrometeoroid density near the Sun may be considerably less than near Earth, some micrometeoroid protection of the sail will probably be necessary after it departs the occulter. No spacecraft has ever performed a close flyby of the Sun, so knowledge of the solar corona through which the solar sail starship must fly is imperfect. But the solar magnetic field there is perhaps 20 to 30 times stronger than the field at the surface of Earth.

The Future of Starsails

In 1974 at the Jet Propulsion Laboratory, a flurry of activity began that may be auspicious for the future of solar sail technology applied to starflight. Space scientists at JPL realized that a solar sail would be the ideal propulsion system to transport a payload of instruments to rendezvous with Halley's comet when it returned to the inner Solar System in 1985–1986. The mission was never flown because a different type of propulsion system— a solar electric rocket or ion engine—was voted less chancy, but more fundamentally because the beleaguered U.S. space program ultimately had no money for a mission to Halley's comet (16). But the seeds of serious solar sail development were planted.

A specific design for the Halley solar sail was chosen, the invention of engineer Richard MacNeal. It was to be a 12-vaned, windmill-like structure called a "heliogyro" that would stiffen by centrifugal force as sunlight caused it to rotate. It was to be launched in 1981 or 1982, building up adequate speed to match the comet's motion years later so that the spacecraft could fly formation with the comet perhaps for months and make detailed studies of its nucleus.

Following the defunct heliogyro system were the plans of a private group, the Pasadena-based World Space Foundation, to deploy an experimental solar sail 100 feet on edge from a future space shuttle payload. The group built a half-scale version of this sail in 1981 and has been waiting for a flight opportunity for their larger completed prototype (17).

Perhaps these early solar sail efforts have begun a technological evolution that will lead in the next century to serious consideration of solar sail starflight. The prospect of launching solar sail starprobes at velocities as high as 0.01 c is an intriguing possibility. Although peak velocities for solar sail robot probes may never exceed 10% of fusion rocket probes, the cost of solar sail starships could be orders of magnitude less than these more energetic craft. The reliability requirements for a centuries-long solar sail starprobe are certainly more stringent than for the multidecade flight of a fusion rocket, but at least the sail itself might well be more reliable than a fusion engine.

Missions by solar sail starship habitats lasting millennia may never be attractive to terrestrials. However, long-term residents of space colonies in the solar system or members of other spacefaring civilizations in the galaxy may not feel equally constrained. Furthermore, as we have seen, there may be undiscovered stars much closer than Alpha Centauri, so we may eventually discover destinations only a few centuries away via inhabited solar sail starships. So in addition to research on solar sail materials, starship dynamics, and logistics, it is not too soon to undertake sociological studies of the problems of maintaining interstellar colonies. Whether clipper ships of the galaxy will ply the interstellar ocean in the twenty-first century, as white-sailed Yankee clippers once did in the nineteenth, is a question our children may be able to answer.

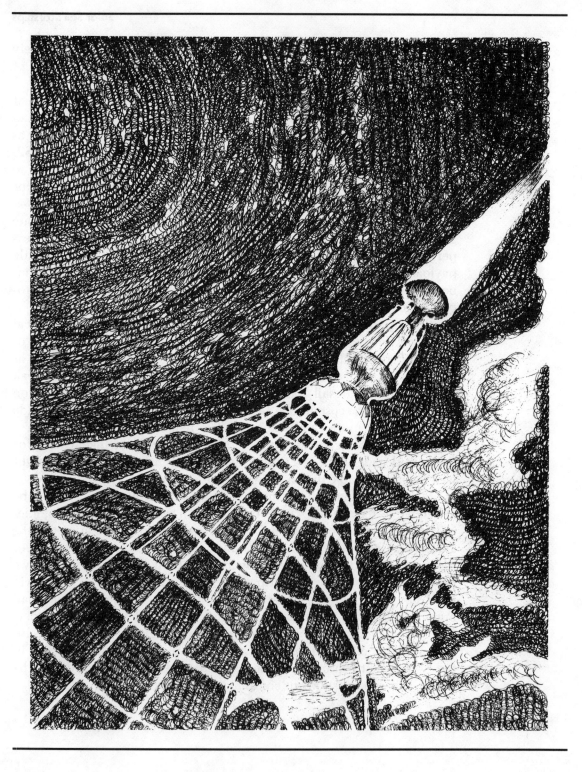

7 Fusion Ramjets

Hitch your wagon to a star.

Ralph Waldo Emerson,
1803–1882,
"Society and Solitude"

Hold the lighted lamp on high,
Be a star in someone's sky.

Henry Burton, 1840–1930,
Pass It On

Bootstrapping Starflight

Imagine a starship so fast that it could travel to any point in the visible universe in mere decades of shipboard time. Moreover, the craft would not require an extensive onboard fuel supply because it would gather propellant and its store of energy from the surroundings, powered and propelled by the most common substance in the cosmos: ubiquitous interstellar hydrogen. The fusing of protons to form helium within the craft's mighty engines would release enormous energy and create a powerful exhaust. This may well be the way to get "something for nothing" and pull a starship across space by the "bootstrap" of the universe, the tenuous interstellar medium.

To envision the extraordinary craft is to recapture the excitement in astronautical circles in the early 1960s, when the first technical papers about the interstellar fusion ramjet appeared. As word of it spread, the interstellar ramjet also galvanized the science-fiction community. If the concept were ever reduced to hardware, the galaxy might brim not only with the likes of Klingons or Romulans, but perhaps with separately launched expeditions by Serbo-Croatians, Tahitians, Hondurans, or Icelanders. After a tiny blip in cosmic time, the numerous independent cultures of the Solar System would blend into the galaxy-wide cosmic culture.

But early optimism has been tempered by further assessments of the far-reaching proposal. Instead of accelerating to near lightspeed in about a year of ship time, fusion ramjets that might be realized seem limited to

10 or 20% of the speed of light. And a fusion ramjet starship would most likely require an initial charge of thermonuclear fuel from the Solar System because it may be well nigh impossible for human technology to duplicate the proton–proton (p–p) fusion reactions that occur deep within the cores of main sequence stars.

But even if we never build a proton–proton fusion ramjet, the effort spent investigating it will not have been wasted. Researchers have examined a number of the concept's derivatives, including the Ram-Augmented Interstellar Rocket (RAIR), the solar- or laser-electric interstellar ramjet, and the solar-electric interplanetary ramjet. Ramjet "runways" are a possibility: fusable isotopes of hydrogen and helium predeployed in the pathway of an accelerating ramjet. And there may be other applications: Orbiting the Sun or another star, a ramjet's electric or magnetic ramscoop could gather fusion fuel from the solar wind plasma; and deflecting interstellar ions with a magnetic or electric ramscoop field would be an ideal way to decelerate an interstellar craft at journey's end.

As fantastic as these ideas may seem, we should be open-minded about them, not cavalierly rule them out because of their current infeasibility. After all, the march of technology is full of well-known surprises and serendipity. To cite one example: the laser was invented virtually at the same time that the fusion ramjet was proposed. Lasers have since become the most promising way, in theory, to ionize the advancing path of a fusion ramjet and facilitate electromagnetic fuel collection.

The Bussard Ramjet

In 1960 Robert Bussard published his seminal article on the interstellar fusion ramjet (1). While at Los Alamos Scientific Laboratory, he conceived this distant relation to chemical combustion ramjets that operate in Earth's atmosphere. Bussard's interstellar fusion ramjet would use a *ramscoop* to funnel charged particles or ions from a wide cross-sectional area of onrushing interstellar medium into the ship's fusion reactor. The scoop would consist of an expansive magnetic or electric field (Figure 7.1).

Its fusion engine would use nuclear reactions that occur in Sunlike stars—converting hydrogen directly into helium—rather than reactions of heavier nuclei that are now being tested in prototypes of fusion power reactors or in thermonuclear weapons (see Technical Note 7–1). The energy liberated in the fusion reaction would then accelerate the reaction products rearward, producing thrust. The dynamics of the fusion ramjet, treated nonrelativistically, are outlined in Technical Note 7–2.

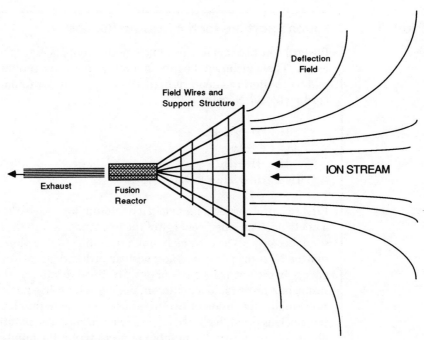

Deflection
Field

Field Wires and
Support Structure

ION STREAM

Exhaust Fusion
Reactor

Figure 7.1 The Bussard ramjet.

Bussard did not attempt to design the ramscoop in great detail, but he did estimate the ramjet's performance in an ionized, high-density region of the interstellar medium. He found that for a starship mass of 1000 tons, an interstellar proton density of $10^9/m^3$, and a 100% efficient hydrogen fusion engine, the craft could accelerate almost indefinitely at one g! Setting out with a very low initial velocity—a few tens of km per second—the craft would approach light velocity within a year.

These ideas were evolving during the early days of the space age, a time when the frontier of exploration was rapidly expanding. It was not long before the popular press got word of the ramjet. In 1964, *New York Times* science writer Walter Sullivan's acclaimed book, *We Are Not Alone,* appeared, a popular treatment of the prospects for contact with extraterrestrial civilizations (2). The work focused on radio communication; though departing from technical orthodoxy, Sullivan described several methods of *direct* contact between galactic civilizations, and prominent among them was the interstellar ramjet. Also that year, Stephen Dole and Isaac Asimov published *Planets for Man*, the popular version of a Rand Corporation study of possible habitable worlds within 21.4 ly of the Sun (3). They cited Bussard's ramjet in their discussion of possible ways to reach these hypothetical oases in space.

Fusion Reactions for the Bussard Ramjet

Bussard considered a few possible reaction sequences for the interstellar ramjet (1). These included the proton–proton (p–p), the proton–deuterium (p–d), and the deuterium–deuterium (d–d) reactions. The proton–proton chain, according to Eva Novotny (26) is:

$$^1H + {}^1H \rightarrow {}^2H + e^+ + v$$
$$e^+ + e^- \rightarrow \gamma$$
$$H + {}^2H \rightarrow {}^3He + \gamma$$
$$^3He + {}^3He \rightarrow {}^4He + 2\ {}^1H$$

First two protons react and yield a deuteron, an antielectron or positron, and a neutrino. The kinetic energy of the neutrino is lost to the fusion reactor (or star), because of the extremely small reaction cross-section of neutrinos with matter. Next the positron reacts with an ordinary negative electron, yielding energy in the form of a gamma ray. The third reaction in the sequence is the fusion of a proton and a deuteron, which yields a ^3He nucleus and a gamma ray. Finally, the fusion of two ^3He nuclei completes the chain, resulting in a ^4He nucleus (or alpha particle) and two protons. The notation used in the above sequence gives the number of protons plus the number of neutrons in the nucleus as a superscript to the left of the element symbol. In the p–p chain, 0.007 of the initial reactant mass is converted into energy. Neglecting the energy lost in the neutrino, 26.20 Mev of energy is released. The proton–deuteron reaction considered by Bussard is the third reaction in the above chain.

In the range of temperatures and pressures for which present fusion experiments are striving, a deuterium–deuterium reaction has two possible outcomes—either a ^3He nucleus and a neutron or a ^3He nucleus and a proton:

$$^2H + {}^2H \rightarrow {}^3He + n$$
$$\rightarrow {}^3He + {}^1H$$

In the p–d and d–d reactions, the fraction of reactant mass converted into energy are respectively 0.002 and 0.001. In practical reactors, much of the electrically neutral neutron's energy might be lost.

Bussard compared the cross-sections of the three reaction sequences over a wide range of particle kinetic energy. The d–d reaction cross-section is 10^{20}–10^{24} times greater than that of the p–p reaction. The cross-section of the p–d reaction is 10^{16}–10^{20} times greater than that of the p–p reaction. In the energy range of tens of Kev, the d–d and p–d reactions require respectively only 10^{-12} and 10^{-8} of the p–p reaction plasma density. Whitmire estimated that the characteristic dimension of a p–p reactor would be about 7000 km—formidable even for a million kilogram starship!

Fusion Ramjet Dynamics (Nonrelativistic)

Consider a ramjet with mass M_s, velocity V_s, moving through an interstellar medium of ion density, ρ. A is the effective intake area of the ramscoop and the average mass of a scooped-up interstellar ion is m_i. In the fusion reactor, fraction, ϵ, of the reaction mass is converted into exhaust kinetic energy. Consider the momentum of the ship and a fuel parcel in the interstellar medium:

$$M_s \dot{V}_s = \dot{M}_f V_e$$

where $\dot{V}_s$ is ship acceleration, $\dot{M}_f$ is fuel mass collected per second and V_e is the exhaust velocity relative to the interstellar medium. From the equation of mass conservation,

$$M_f = A\rho m_i V_s$$

To determine V_e, compare the kinetic energy of the fuel before and after it reacts in the fusion engine. Mass is converted into fusion product kinetic energy as:

$$\epsilon \dot{M}_f c^2$$

where c is the speed of light. The total energy is conserved, so comparing the before and after kinetic energies, get:

$$v_e^2 + 2V_s V_e - 2\epsilon c^2 = 0$$

Applying the quadratic formula, finding the physically meaningful solution, and selecting $V_s^2 > 2\epsilon c^2$, get:

$$V_e = \epsilon \frac{c^2}{V_s}$$

Substituting above, get:

$$\dot{V}_s = A\rho m_i \epsilon \frac{c^2}{M_s}$$

Observe that ramjet acceleration is independent of spacecraft velocity!
As an example, consider a scoop diameter of 2000 km, $M_s = 10^6$ kg, $\epsilon = 10^{-3}$, $\rho = 10^{-6}/m^3$, $m_i = 1.67 \times 10^{-27}$ kg (protons). Putting these values into the equations, $\dot{V}_s = 0.5$ m/sec^2, or about 0.05 g. The starship would reach about half the speed of light after accelerating for 10 years.

Astronomer Carl Sagan, whose name has long been associated with the search for radio signals from extraterrestrial civilizations, inadvertently may have done more than anyone to make the potential of the interstellar ramjet known. In his thought-provoking technical article, "Direct Contact Among Galactic Civilizations by Relativistic Spaceflight," Sagan considered a 1000 ton fusion ramjet that would accelerate at one g in a typical interstellar region having a proton density of $10^6/m^3$ (4). He assumed the diameter of the intake zone to be about 4000 km, much larger than the 160 km required in the thousand-times less dense medium supposed by Bussard.

Examining the relativistic flight kinematics of a ship accelerating at 1 g, Sagan found that after three years of shipboard time, the craft would pass Alpha Centauri; within 4 shipboard years, it would pass Epsilon Eridani at 11 ly; after 11 years of ship time, it would reach the Pleiades star cluster(M45) 400 light years away. Leaving the Milky Way galaxy far behind, ramjet adventurers would pass the nearby irregular galaxies—the Magellanic Clouds—after only 23 years, and would reach M31, the nearby Great Galaxy in Andromeda, in 25 years. Sagan found that a 1 g starship could, in theory, reach the outskirts of the visible universe within the lifetime of a crew member. Returning to Earth, and no doubt anticipating fat pension checks, the crew would instead find an aging white dwarf star in place of the once vibrant young Sun. Because of relativistic time dilation, billions of years would have passed on the unaccelerated Earth. (The reality of this supposed "paradox" of relativity is quite secure. See Appendix 5.) In light of such possible advanced technology, Sagan estimated that thousands of alien cultures may have visited the Solar System during past eons. He went so far as to recommend a search of the Solar System for possible alien bases, as well as a careful study of ancient legends that might represent "contact myths."

Few would have known about these speculations if Carl Sagan and Soviet astrophysicist I.S. Shklovskii had not included them in their immensely successful collaboration, *Intelligent Life in the Universe* (5). The fusion ramjet was suddenly virtually synonymous with starflight—the favored mode of interstellar transport. Science fictional ramjets appeared in the work of Larry Niven, *A Gift from Earth*, and Poul Anderson related the haunting circumuniverse flight of the ramjet Leonora Christine in *Tau Zero* (6,7).

But while enthusiasts trumpeted the idea, physicists and engineers were beginning to explore the practical problems in the design of a technologically feasible ramjet. John Fishback of MIT considered the possible physical limitations on ramjet performance, basing his arguments on the structural strength of materials. He estimated the maximum range of an ideal fusion ramjet built of diamond would be about 10,000 ly for missions still within a crew member's lifetime (8). The distance is very far to be sure, but far from the "edge of the visible universe." British researcher Anthony Martin considered the electric or magnetic field of the ramscoop as a deflector of ionized interstellar atoms that could decelerate a speeding starship (9,10).

The *reaction cross-section* of the proton–proton fusion reaction sequence is orders of magnitude lower than that of the deuterium–deuterium fusion reaction, which might be feasible early in the twenty-first century, as outlined in Technical Note 7–1. The p–p fusion reaction is therefore very far from technological feasibility, requiring the kind of temperatures and pressures that perhaps only the cores of stars can provide. For this and other reasons, some researchers, notably Freeman Dyson and Thomas Heppenheimer, have said that the fusion ramjet may be impossible (11,12). The indefatigable Robert L. Forward, however, believes that the ramjet concept is too potentially valuable to be discarded despite its difficulties (13). Undaunted, some researchers have suggested alternatives to the proton–proton fusion ramjet.

The Catalytic Ramjet

One alternative is the catalytic nuclear ramjet of Daniel Whitmire (14). The Sun fuses protons (hydrogen nuclei) in its core with a temperature reckoned in millions of degrees, but its surface temperature is only about 6000°K. Many stars burn much hotter than the Sun and thus consume their fusion fuel much faster. The thermonuclear reactions within these hot stars are *catalytic:* a particular isotope, which is not used up in the reaction, is employed to speed the conversion of hydrogen atoms to helium and released energy. Two known fusion catalysts are the isotopes of carbon and neon, ^{12}C and ^{20}Ne. In this nuclear notation, 12 and 20 refer to the total number of nucleons (protons plus neutrons) in the nucleus of the atom.

Though Technical Note 7–3 gives these reaction cycles in more detail, catalytic thermonuclear fusion is understandable from the following "net" equations for uncatalyzed and catalyzed fusion:

$$4\ ^1H \rightarrow\ ^4He + Energy \qquad \text{(proton–proton uncatalyzed)}$$
$$4\ ^1H +\ ^{12}C \rightarrow\ ^4He +\ ^{12}C + energy \qquad \text{(carbon catalyzed)}$$
$$4\ ^1H +\ ^{20}Ne \rightarrow\ ^4He +\ ^{20}Ne + energy \qquad \text{(neon catalyzed)}$$

Catalytic Ramjet Reaction Sequences

Possible reaction sequences for Whitmire's catalytic ramjet include the "CNO catalytic Bi-Cycle" and the "Ne–Na chain":

CNO Cycle
$$^{12}C + {}^1H \rightarrow {}^{13}N + \gamma$$
$$^{13}N + {}^1H \rightarrow {}^{14}O + \gamma$$
$$^{14}O \rightarrow {}^{14}N + e^+ + \nu$$
$$^{14}N + {}^1H \rightarrow {}^{15}O + \gamma$$
$$^{15}O \rightarrow {}^{15}N + e^+ + \nu$$
$$^{15}N + {}^1H \rightarrow {}^{12}C + {}^4He$$

Ne–Na Chain
$$^{20}Ne + {}^1H \rightarrow {}^{21}Na + \gamma$$
$$^{21}Na \rightarrow {}^{21}Ne + e^+ + \nu$$
$$^{21}Ne + {}^1H \rightarrow {}^{22}Na + \gamma$$
$$^{22}Na + {}^1H \rightarrow {}^{23}Mg + \gamma$$
$$^{23}Mg \rightarrow {}^{23}Na + e^+ + \nu$$
$$^{23}Na + {}^1H \rightarrow {}^{20}Ne + {}^4He$$

As with the p–p reaction, the positrons (e^+) in both chains will react with electrons, liberating additional energy as gamma rays. Note that in both chains, the catalyst (^{12}C or ^{20}Ne) is conserved and the net effect is to convert four protons to one helium nucleus.

At sufficiently high energies (temperatures), the slowest reactions in the carbon and neon cycles have reaction rates 10^{18} to 10^{19} times greater than those of the uncatalyzed proton–proton fusion reaction. Yet these reactions will still be about a million times slower that the deuterium–deuterium reaction, which will be of practical utility within a few decades.

Whitmire estimated some of the parameters of a hypothetical catalytic fusion reactor that could be used in an interstellar fusion ramjet. For a 1000-ton starship accelerating at a steady 1 g with a 100% efficient reactor, the reactor power output is 10^{11} megawatts, about 10,000 or more times the energy generated by present-day global civilization! (See Appendix 6.) The ion temperature in the reactor is $10^9 °K$ (corresponding to an ion energy of 86.2KeV). The radial dimensions of the cylindrical reactors for the carbon and neon cycles are respectively 19 and 9.6 meters. The magnetic field necessary to confine the plasma is 2×10^7 Gauss, about 100 times greater than the most powerful magnetic fields generated with current technology. Because of the extreme requirements in the design of a catalytic fusion reactor operated in the steady state, Whitmire has suggested incorporating micropellet pulsed-fusion or the more speculative positive-ion Migma reactor (14,15).

But how to collect the interstellar fusion fuel? Sagan had suggested a ramscoop based on the idea of a superconducting *magnetic flux pump.* But such a device may prove very difficult to implement, certainly in interstellar regions of low- to moderate-mass density. To get around this problem, Whitmire suggested combining electric and magnetic fields to collect interstellar ions (see Chapter 8). In largely non-ionized or neutral regions of the interstellar medium, a starship might project laser beams ahead of it to ionize hydrogen atoms, thus greatly enhancing the efficiency of matter collection by the ramscoop's magnetic or electric fields (16). Whitmire and Matloff and Fennelly have analyzed the fruitful concept (see Technical Note 7–4). Whatever their design, feasible electromagnetic ramscoops would likely be limited to low accelerations of 10^{-4} to 10^{-3} g (see Chapter 8).

Technical Note 7–4

Ionizing Interstellar Hydrogen by Laser Beam

Author Matloff and A. Fennelly (16) long ago investigated the vacuum ultraviolet laser, projected ahead of a ramjet starship, as a way to increase the level of ionization of hydrogen in the ICM. Ideally, the laser wavelength should vary as the ship's velocity changes in order to compensate for the frequency change due to the Doppler effect. Light of $0.0916\mu m$ wavelength has just enough energy to ionize ground-state hydrogen.

By considering the volume traversed by a laser photon (π L $\lambda^2/4$, where L = beam length and λ = wavelength), the total volume of the entire laser beam ($\pi R^2 L$, where R = beam radius), and the definition of photon energy (E = h c / λ, where h is Planck's constant), the laser energy for 100% ionization is E = $4hcR^2/\lambda^2$. If R = 50,000 km, E = 2×10^{12} Joules.

If the laser were turned on for 50 days and the pulse repeated every 230 days, the laser power is a modest 5×10^5 watt. Because light travels 1.3×10^{12}km in 50 days, the necessary beam dispersion is 3×10^{-8} radian. This could be achieved if the laser were put at the focus of a 3.8 meter diameter, diffraction-limited Cassegrain telescope.

Although Whitmire agrees that ICM ionization by a UV laser is technologically feasible, he believes that the temperature of the ionized region would reach about 10,000°K. If this were to happen, ionizing by laser would only be feasible for much higher particle densities than are found in the ICM. It will be essential to do much more research on ionizing the ramjet medium, including other methods, such as the high-energy electron beams suggested by Karlovitz and Lewis (17).

Other types of catalytic nuclear reactions may help to reduce the difficult requirements of the fusion ramjet. But a related and less technologically demanding system than the pure fusion ramjet may be the *ducted rocket* or *Ram Augmented Interstellar Rocket (RAIR)*, in which interstellar hydrogen serves mainly to augment the reaction mass rather than to act as fusion fuel. So RAIR is, indeed, an *augmented* rocket. Although more sluggish than the pure ramjet, RAIR requires less fusion fuel than a rocket to reach a given cruise velocity.

RAIR: The Ram Augmented Interstellar Rocket

Although Bussard mentioned the idea of RAIR in his 1960 ramjet paper, British engineer Alan Bond first examined it in detail in 1974. Shown schematically in Figure 7.2, RAIR incorporates two propellant streams. Interstellar hydrogen is still collected by a ramscoop similar to that of the fusion ramjet, but the hydrogen is not converted to helium in a fusion reaction. The incoming hydrogen is instead accelerated with the energy of fusion reactions produced by fuel carried by the starship. An early proposed version of RAIR would use deuterium alone, deuterium–tritium, or deuterium–helium-3 fuel in fusion reactions that even now are close to being realized in practice. A more advanced RAIR might employ the more demanding proton–lithium (^{7}Li) or proton–boron (^{11}B) reactions. Bond suggested diverting a small fraction of the incoming proton stream to strike micropellets of lithium and thereby induce thermonuclear reac-

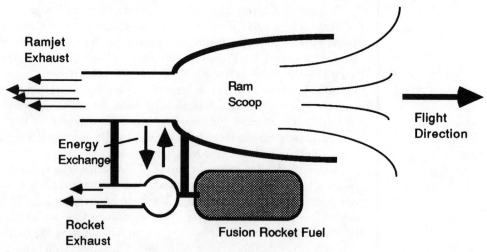

Figure 7.2 Ram-augmented interstellar rocket (RAIR).

Simplified RAIR Kinematics

The kinematics of RAIR are very complex, even treated nonrelativistically, because there are two exhaust streams—one from the fusion reactor and one from the accelerated interstellar reaction mass. As Powell has said, it is unlikely that the two exhaust streams can be mixed (19,20).

If we assume, however, as did author Matloff, that the two exhaust streams can be treated mathematically as well-mixed, we can compare RAIR ("ra" subscript) and rocket ("ro" subscript) performance (29). ϵ is the energy per unit mass of fuel in the fusion reaction *times* the fraction of fuel mass "burned." The exhaust velocities of RAIR and rocket are:

$$V_{e-ra} = c \sqrt{\frac{2\epsilon}{(1+f)}}$$

$$V_{e-ro} = c\sqrt{2\epsilon}$$

where f = interstellar reaction fuel mass/ fusion fuel mass.

Since the fusion reactor burns a total fuel mass M_f, the momentum increments during the same time interval of RAIR and rocket operation are:

$$\Delta p_{ro} = M_f V_{e-ro} = M_f c\sqrt{2\epsilon}$$

$$\Delta p_{ra} = M_f(1+f)V_{e-ra} = M_f c\sqrt{2\epsilon(1+f)}$$

Therefore,

$$\frac{V_{ra}}{V_{ro}} = \sqrt{1+f}$$

According to this approximation, the velocity of a RAIR at burnout will be twice rocket burnout velocity for f = 3 and *three times* rocket burnout velocity for f = 8.

tions. Winterberg suggested boron as an alternative to lithium. Technical Note 7–5 presents a few fundamentals of the kinematics of RAIR, showing in particular that the system is more efficient as the amount of reaction mass goes up. Possible reactions that an advanced RAIR might use are shown in Technical Note 7–6.

Bond compared the performance of RAIRs of varying efficiency with rockets; his analysis uses starship mass ratio as the performance param-

Nuclear Reactions for RAIR

Alan Bond suggested that while reacting ^{3}He-deuterium or deuterium–deuterium fusion fuel in isolation from the interstellar propellant stream, some interstellar protons might be siphoned off to cause fusion reactions in lithium fuel pellets:

$$^1H + {}^7Li \rightarrow 2\ {}^4He$$

with the fraction 0.0023 of the reactant mass converted to energy. A pioneer in the field of staged fusion microexplosions, F. Winterberg has suggested an alternative to the lithium–proton reaction—the proton–boron fusion/fission reaction:

$$^1H + {}^{11}B \rightarrow 3\ {}^4He$$

Both boron and lithium may be accessible extraterrestrial resources from the Moon or asteroids. All the products of these reactions are charged particles, so no energy will be lost to neutrinos or neutrons which can't be directed into the exhaust jet.

---

eter (18). Conclusion: Even an inefficient RAIR requires much less fuel than a comparable fusion rocket. Conley Powell and A.A. Jackson IV have carried out similar analyses, but have incorporated the correct relativistic dynamics (19–22).

It is useful to compare low-speed travel with RAIR with the flight of a pulsed fusion rocket, such as the Project Daedalus starship (Chapter 4). The Daedalus pulsed thermonuclear engine would have an exhaust velocity of 0.03 c and would require a mass ratio of 2.7 to achieve a cruise velocity of 0.03 c. By way of comparison, Powell's calculations for low-velocity flight with RAIR indicate that a similar mass ratio allows RAIR to cruise at 0.04 to 0.07 c, depending on subsystem efficiencies, the fusion fuel burnup fraction, conversion efficiency to directed exhaust energy, ion collection rate, and so forth (21). With still reasonable mass ratios, RAIR could well reach 0.1 to 0.2 c.

In 1980 the authors considered how an interplanetary civilization might use RAIR to colonize a neighboring solar system (23). We estimated that a 1.8×10^4 ton starship, excluding fuel, would be required to support a 50- to 100-person group for a few centuries of interstellar travel aimed at founding a planet or asteroid-based colony in another solar system. Rather than using a rare isotope like ^{3}He in the fusion engine, lithium or boron would be mined from an asteroid or comet before departure. The analysis of lunar rocks reveals that lithium and boron are reasonably abundant on the Moon and, presumably, in asteroids or comet nuclei as well. Mining about 10^5 tons of hydrogen and lithium or boron would allow RAIR to cruise at about 0.1 c. Deceleration into the destination solar system would be by electrostatic or electromagnetic drag screens (see Chapter 8).

Fuel for the Starship

Interstellar fusion ramjets, RAIR, and pulsed fusion rockets depend critically on finding adequate supplies of fusion fuels within the Solar System or in the interstellar medium, which is predominantly hydrogen. As for fuel from the interstellar medium, although Bussard's original ramjet was to "burn" protons directly, some slower variants of the ramjet would employ interstellar deuterium (^{2}H) and ^{3}He. We can forget about the radioactive isotope tritium (^{3}H) because its decay half-life is about 12 years, and its natural abundance is negligible. Whitmire's catalytic ramjet requires neon, carbon, or nitrogen catalysts, replenishable from the interstellar medium to cover inevitable leakage.

However, the interstellar medium is not the only feeding ground for fuel-hungry starships. A RAIR ship might burn deuterium or ^{3}He extracted from either the solar wind or mined from the atmosphere of Jupiter. More advanced RAIR technology could obtain lithium or boron from comets and asteroids. Thus, the four fuel sources on which fusion ramjet/RAIR technology may depend are the interstellar medium, the solar wind, giant-planet atmospheres, and asteroids, or comets:

The Interstellar Medium and Solar Wind

The interstellar medium is not nearly as uniform as one might suppose. In an interstellar gas cloud or nebula, the hydrogen density might be as high as 10^9 protons/m^3. If no hot stars are nearby, the gas in the nebula—a region of neutral gas called a *HI region*—will not be ionized. An ionized cloud with a similar gas density is an *HII region*, its hydrogen being substantially ionized because of the proximity of one or more hot stars. The Solar System does not reside in either an HI or HII region but is within the so-called *Intercloud Medium* or *ICM*. Ultraviolet light measurements from Earth-orbiting observatories have suggested a density of neutral hydrogen within the ICM of 2×10^5 to 3×10^5 /m³ and an ICM proton density of 5×10^4/m³. The "average" interstellar medium, which Sagan used in his analysis of the ramjet, comes from "smoothing-out" observations of the ICM, HI, and HII regions over a 1000 light year path.

The cosmic abundance ratios of some of the fusion fuels of interest are:

Deuterium/hydrogen—1.4×10^{-5}
Helium-3/hydrogen —3×10^{-5}
Carbon/hydrogen —3.5×10^{-4}
Nitrogen/hydrogen —8.5×10^{-5}
Neon/hydrogen —7.6×10^{-5}

These ratios should be roughly independent of the type of interstellar region. Chapter 11 presents a more complete discussion of the interstellar medium.

Meeting the interstellar medium at the heliopause is the *solar wind*—a stream of ions which the Sun emits, as do presumably other stars. At the orbit of the Earth, the velocity of the solar wind is typically 400 km/sec. The ion density ranges from 2 to 10 ions/m³. While the solar wind intensity varies with the solar activity cycle, the forementioned cosmic abundance ratios should prevail on average.

Atmospheres of the Giant Planets

In the Project Daedalus study, the research team suggested obtaining ³He for the starship's pulsed fusion reactor from the atmosphere of Jupiter using large balloons floating in the planet's atmosphere. The study group estimated pertinent abundance ratio for the Jovian atmosphere: ³He/H = 1.7×10^{-5}.

Asteroids and Comet Nuclei

In rocky asteroids, the fractional abundances of ⁷Li and ¹¹B found on the Moon would likely prevail. Except in the surface layers of asteroids (and the Moon), which are continuously exposed to the solar wind, hydrogen, deuterium, and ³He should be extremely rare. The fractional abundances of ^{7}Li and ^{11}B to be expected are 10^{-5} and 3×10^{-6} respectively. Now consider a spherical asteroid 2 km in diameter, with a density of 300 kg/m^3. It would contain about 10^5 tons of ^{7}Li and 3×10^4 tons of ^{11}B. Of course, surface samples of asteroids—much less comets—have yet to be returned to Earth, but if a comet nucleus is modeled as layers of ice deposited on a rocky core, perhaps deuterium could be obtained from the ice layers and lithium and boron from the core.

8 Interstellar Ion Scoops

Let us create vessels and sails adjusted to the heavenly ether, and there will be plenty of people unafraid of the empty wastes.

Johannes Kepler, in a letter to Galileo, 1610

Over all the sky—the sky! far, far out
of reach, studded, breaking out,
the eternal stars.

Walt Whitman, 1819–1892,
Leaves of Grass

Gathering Almost Nothing

Observing the flight of a ramjet or RAIR starship, one watches a grotesquely distorted cosmic "fish," almost all mouth and very little stomach, gulping its way across the interstellar ocean. The looker would marvel at the workings of that cavernous yet delicate mouth and how it can sate the voracious appetite of its tiny, by comparison, fusion machine body. To gather what is—after all—almost nothing is really something!

Early investigators of the interstellar ramjet, such as Bussard and Sagan, did not minimize the difficulty of gathering interstellar ions efficiently over a fantastically large radius, but their technical papers did not attempt to design a ramscoop in even minimal detail. They merely estimated that the radius of a ramscoop would have to be hundreds or thousands of kilometers, depending on the density of the medium through which the starship would travel. Bussard suggested that the scoop could be either a magnetic or an electrostatic field; Sagan proposed a superconducting magnetic flux pump; and Fishback called for a slowly varying or a static magnetic field (1–3).

Where would such ethereal scoops "leave off" and the unaffected interstellar medium begin? Anthony Martin suggested defining the intake zone of a magnetic ramscoop as the region for which $B > B_G$, where B is the field strength of the scoop and B_G is the magnetic field in the local interstellar medium, predominantly the galactic field (4; see Figure 8.1). (This criterion unquestionably deserves further theoretical attention, particularly for high-speed ramjet flight.) The galactic magnetic

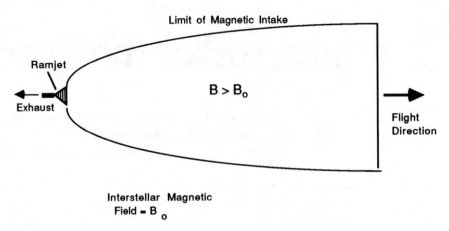

Figure 8.1 The magnetic ramscoop.

field near the Solar System is very weak—thought to be about 10^{-6} gauss, a microgauss. By comparison, the strength of the geomagnetic field at Earth's surface is on the order of 0.5 gauss. Martin also proposed a superconducting *solenoid* (coil) to generate the ramscoop's field. But the first (and to our knowledge *only*) attempt so far to design a superconducting solenoid scoop was carried out by author Matloff and Alphonsus Fennelly in 1974 (5).

A Superconducting Magnetic Ion Scoop

Superconductivity is the vanishing of electrical resistance, a property that in both low- and high-temperature superconductivity depends on complex quantum mechanical phenomena. It is possible to sustain a large electrical current in a superconducting circuit with no continuing power input. Hence, the opportunity with superconducting solenoids to create enduring, high magnetic fields. Until the recent spectacular breakthroughs in the field of high-temperature superconductivity ("high T_c," or high critical temperature) beginning in 1986, superconductivity was confined to temperatures ($\approx 20°K$) not much above absolute zero. Now, in laboratories around the world—including many high-school labs!—researchers routinely observe superconductivity in certain ceramic materials at temperatures above $77°K$, the "balmy" temperature of cheap liquid nitrogen at normal pressure. (The highest T_c yet confirmed is about $125°K$.)

With prospects improving for finding superconductors that will work at "room" temperature and perhaps even above, the future looks bright for many new applications, not the least of which might be magnets for

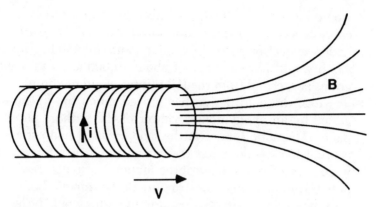

Figure 8.2 Superconducting ion scoop.

fusion reactors and other gadgets for starflight. In the Matloff-Fennelly superconducting ion scoop (Figure 8.2), magnetic field lines emerge from a thin-film, curved cylindrical solenoid. Current flows in a thin film (10^{-6}m) layer of conventional superconducting alloy, Niobium-Tin (Nb_3Sn). The positive interstellar ions caught in the field tend to gyrate around the magnetic field lines, as shown in the figure. The scoop is curved for reasons relating to *Silsbee's rule*, which defines the maximum supercurrent that can be supported in a wire. Large currents of 10^{11} amps could be supported in a curved cylindrical scoop of length 0.4 km.

Because an unsupported thin-film scoop would collapse under even very low accelerations, Matloff and Fennelly assumed a 5×10^{-6}m thick copper substrate and the supporting structure shown in Figure 8.3. They calculated that a total scoop plus support structure mass of 100 tons

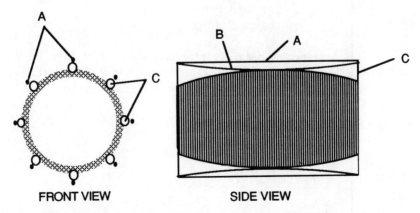

FRONT VIEW   SIDE VIEW

Figure 8.3 Ion scoop and support structure. (a) Back beams. (b) Curvature support. (c) Supporting end-caps.

would be adequate for spacecraft accelerations less than 0.04 g. Technical Note 8–1 presents the equations for the effective area of the scoop and its ion collection rate. In a microgauss interstellar magnetic field, the scoop would have a range of about 10,000 km and would collect a few grams per second from the ICM (inter-cloud medium).

Alas, the bold design has proved much too optimistic, as was suggested by an anonymous reviewer of the authors' subsequent technical paper. Magnetic forces on the scoop, called hoop stresses, would likely cause it to explode unless the current (and therefore the magnetic field) were reduced or the mass of the supporting structure were increased. But even if the entire massive structure of the starship body were used to support the scoop, its effective maximum radius would be less than 1000 km.

Technical Note 8–1

Effective Field Radius and Limitations of a Magnetic Scoop

Matloff and Fennelly described the effective field radius of their superconducting solenoid ion scoop by first assuming that the scoop field, B, would be effective in gathering interstellar ions when $B > B_G$, the local galactic magnetic field (5). According to A. Dalgarno and R. A. McCray, B_G is approximately 10^{-6} gauss or 10^{-10} weber/m² (18). The dipole strength, M, of the solenoid is:

$$M = \frac{i\pi r^2}{4\pi\epsilon_0 c^2}$$

where i = superconducting current
r = physical radius of the scoop solenoid
ϵ_0 = permittivity of free space
c = the speed of light

At transverse distances, t (perpendicular to the direction of starflight), and longitudinal distances, z (in the direction of starflight), the transverse and longitudinal components of the scoop field will be:

$$B_t = \frac{2M\sin(\theta)\cos(\theta)}{(z^2 + t^2)^{3/2}}$$

$$B_z = \frac{2M\cos(\theta)^2}{(z^2 + t^2)^{3/2}}$$

where $\tan(\theta) = t/z$.

Therefore, the condition for an ion to be drawn in is:

$$\sqrt{B_t^2 + B_z^2} \geq 10^{-6} \text{ gauss}$$

For a scoop with a 0.4 km physical radius and 0.4 km long, a 10^{11} amp superconducting current is possible. The scoop effective field radius is 10^4 km.

However, as discussed by Matloff and Fennelly, a limitation on scoop performance is the magnetic pressure on the solenoid walls (9):

$$P_m = \frac{B_s^2}{2\mu_0}$$

where B_s is the solenoid magnetic flux density and μ_0 is the permeability of free space.

Considering the strength limitations on the solenoid structure, Matloff and Fennelly next related the solenoid wall thickness, t_w, necessary to keep the scoop from exploding, to the maximum permissible hoop stress on the solenoid, σ_h, and the radius of the solenoid:

$$t_w = \frac{B_s^2 r}{2\mu_0 \sigma_h}$$

Using values of σ_h for existing and projected engineering materials and values of t_w based on structural supporting mass in the range 2×10^7 to 5×10^8 kg, one can show that effective scoop radii much larger than 2000 km are unlikely.

The Whitmire Electromagnetic Ramscoop

Whitmire was first to suggest combining electrostatic and magnetic fields to reduce the structural requirements on magnetic scoops, while at the same time increasing a scoop's effective field radius (7). His electric/magnetic scoop design is shown in Figure 8.4, the purpose of the arrangement being to reduce or eliminate *proton drag* on a relativistic catalytic ramjet. Whitmire designed the electrostatically charged grids portrayed in the figure, which have the following effects: In regions A and D, an onrushing proton sees no electric field (E = 0). When it arrives in region B after crossing the forward negatively charged grid, the proton decelerates until it moves with the velocity of the ship. After undergoing thermonuclear fusion in the ship's reactor, the reaction products are

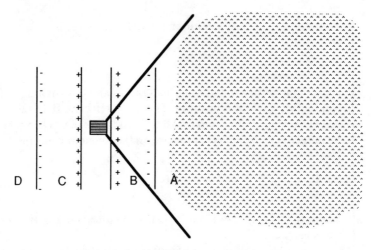

Figure 8.4 Whitmire's electric/magnetic scoop.

expelled through region C. There, the original kinetic energy of the proton fuel relative to the ship, plus the nuclear energy released in the reaction, is transferred to the exhaust. Whitmire claimed that the combined field approach would greatly reduce the mass required to support a magnetic scoop of equivalent intake area, but he did not specify how the extensive grids were to be held in position during high ramjet acceleration.

The Matloff/Fennelly Electromagnetic Ion Scoop

Building on these basic designs, Matloff and Fennelly continued their study of interstellar ion scoops and arrived at two different designs. One of their electric/magnetic scoop configurations appears in Figure 8.5 (8). Interstellar protons are attracted by the leading negative grid while much lighter electrons are repelled. After the protons have gyrated around the magnetic field lines of the solenoid, they would be decelerated by the positively charged spherical surface—similar to Whitmire's approach. Accelerated by positively charged surfaces after emerging from the ramjet reactor or RAIR accelerator, the positive ion exhaust would exit into a region of low or zero electric field. Because of the repulsion of electrons by this scoop, some *electron drag* is entailed, but as Technical Note 8–2 demonstrates, electron drag is negligible for velocities less than about 0.2 c.

Matloff and Fennelly proposed another version of this scoop that may be essentially drag-free (9). A positively charged cable projecting ahead

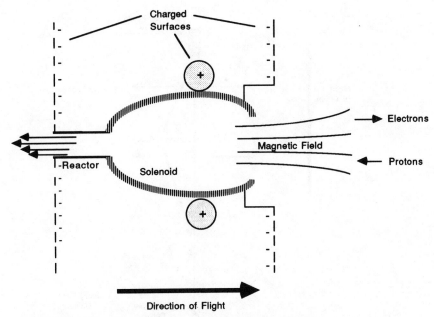

Figure 8.5 Electric/magnetic scoop for ramjet/RAIR.

would repel electrons before they encountered the scoop's main electric field, as shown in Figure 8.6. With a negative charge of 20 coulombs, the electrostatic scoop would have an effective field radius of 3×10^5 km in the ICM. The cable would be about 10^6 km long and its negative charge about 2 coulombs. During acceleration, an auxiliary propulsion unit at the leading end of the cable would keep it deployed (see Technical Note 8–3). To keep the cable in position and prevent it from collapsing, the propulsion unit's acceleration should equal the starship's.

Technical
Note
8–2

Electron Drag and Electric Scoops

We can explore the electron drag of the simple scoop shown in Figure 8.5 following the approach of Matloff (8). A condition for drag-free operation is set by imposing the condition that the magnitude of the exhaust momentum be 10 times the magnitude of the momentum of deflected electrons. Assuming that electrons are deflected at the ship's velocity, $m_p V_p > 10 \, m_e V_s$, where m_p and m_e are proton and electron masses and v_p is the fuel exhaust velocity. If the fuel is exhausted at $v_p = 0.03$ c, the exhaust velocity of the Daedalus starship, electron drag will be insignificant for spacecraft velocities less than about 0.2 c.

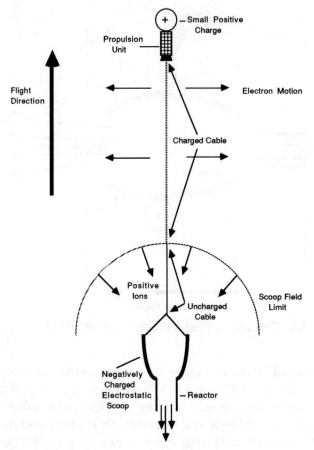

Figure 8.6 Charged cable drag reducer.

Electric Scoop Design Details

As shown by Matloff and Fennelly (9), the mass intake per second by an electric scoop is defined as:

$$\dot{M}_I = \frac{\pi Q_s V_s \rho^{1/3} m_i}{q_i}$$

where Q_s = charge on the scoop
q_i = charge on an interstellar ion
V_s = ship velocity
m_i = mass of an interstellar ion

130

A significant advantage of electrostatic over magnetic collection is the relative insensitivity of $\dot{M}_I$ to variations in ρ.

Electron drag is eliminated with the long charged cable supported by the propulsion unit in front of the effective field limit of the scoop, as shown in Figure 8.6. The small positive charge in front of the propulsion unit protects this system and the cable from proton erosion.

The effective field radius of the scoop is defined as:

$$R_e = \left(\frac{Q_s}{q_i}\right)^{1/2} \rho^{-1/3}$$

the radius for which the scoop electric field is greater than interstellar electric field.

The long charged cable is based upon an earlier design by Forward for Lorentz force turning. Cable length, tensile strength, diameter, density, and mass are respectively: 3×10^6 km, 3×10^8 newton/m², 10^{-5} m, 2000 kg/m³, and 500 kg. According to Forward, a slightly more massive cable could support a 10^4 coulomb charge.

The electric field strength at a radial distance r from the cable is:

$$E_r = \frac{Q_c}{2\pi\epsilon_0 L r}$$

where Q_c is the cable charge and L is the cable length. A negative charge on the cable will deflect interstellar electrons at right angles to the ship's path.

After calculating the electron's potential energy decrease as it travels from cable radial distance r_1 to r_2, the increase in electron kinetic energy is determined from energy conservation considerations. The consequent increase in electronic velocity is written:

$$\Delta V = \ln\left(\frac{r_1}{r_2}\right) \sqrt{\frac{q_e Q_c}{\pi\epsilon_0 L m_e}}$$

where q_e and m_e are electronic charge and electron mass.

For a 10^6 km cable and a 0.04 c spacecraft velocity, 75 seconds are required for an electron to traverse the cable length. Substituting this value of L and assuming $Q_c = 2$ coulombs, we find that the electron velocity increases by 0.011 c as its radial distance from the cable increases by 2.7 times. Matloff and Fennelly concluded that this was adequate to sweep electrons out of the ship's path. So the 3×10^5 km field radius electric scoop would attract protons, but not reflect electrons. Electron drag is therefore minimal.

In electrostatic ion collection, as considered by Matloff and Fennelly, the positive ion fuel would enter the ramjet reactor or RAIR accelerator with an electric field-induced velocity. This would tend to reduce ramjet acceleration below that of a ramjet with a purely magnetic scoop (see Technical Note 8–4). Yet even with the problem of induced fuel velocity, the large electric/magnetic ramscoop would allow a net positive vehicle acceleration *if* the interstellar plasma does not radiate too much energy as it is compressed (an important assumption!). A deuterium/^{3}He ramjet with a mass of 2×10^4 tons could accelerate at 10^{-5} to 10^{-4} g and reach velocities of 0.002 to 0.01 c. The ramjet could conceivably reach Alpha Centauri in about a millennium, but because this is comparable to interstellar solar sailing, we doubt that such a low-speed ramjet will ever be built for interstellar flight.

We hope these early insights on ramscoop dynamics will help other investigators to discover superior designs for starflight propulsion. But devices based on ramscoop technology may have other applications as well, two of particular significance: (1) to gather fuel from the solar wind for other kinds of fusion rockets and (2) as deceleration devices in starflight.

Technical
Note
8–4

Comparing Magnetic and Electric Scoop Efficiencies

Magnetic scoops collect ions without imparting to them a velocity component transverse to the starship. Electric scoops do impart such a velocity. The magnetic scoop is therefore more efficient in RAIR or ramjet applications, as we demonstrate.

From the nonrelativistic treatment in Chapter 7, starship acceleration will be directly proportional to V_e, the exhaust velocity relative to the interstellar medium. This is calculated from the change in fuel kinetic energy ΔKE_f imparted by the accelerator or reactor:

$$\Delta KE_f = 0.5 \Delta m_f [(V_i + V_s + V_e)^2 - (V_i + V_s)^2]$$

where V_i and V_s are the field-induced ion velocity and the starship velocity respectively, and Δm_f is the incremental fuel mass. Rearranging and solving,

$$V_{e \text{ electric}} = -(V_i + V_s) + \sqrt{\frac{(V_i + V_s)^2 + 2\Delta KE_f}{\Delta m_f}}$$

$$V_{e \text{ magnetic}} = -V_s + \sqrt{\frac{V_s^2 + 2\Delta KE_f}{\Delta m_f}}$$

since $V_i = 0$ for a magnetic scoop.

We next define $\Delta V_e = (V_{e\ magnetic}) - (V_{e\ electric})$:

$$\Delta V_e = V_i + [A^2 + B^2] - [C^2 + B^2]$$

where $A = V_s$, $B = [2\Delta KE_f/\Delta m_f]^{0.5}$ and $C = V_i + V_s$.

In the low kinetic energy case, $A > B$ and $C > B$. It is easy to show that

$$\Delta V_e = \frac{(C - A)B^2}{2AC}$$

which is always positive since $C > A$.

In the high kinetic energy case, $B > C$ and $B > A$ and

$$\Delta V_e = V_i + \frac{(A^2 - C^2)}{2B} = V_i \frac{[(B - A) + (B - C)]}{2B}$$

which is always positive because $B > C$ and $B > A$. This means that the exhaust velocity for a magnetic scoop ramjet will always be higher than the exhaust velocity of an electric scoop ramjet. The efficiency of the magnetic scoop ramjet is always higher.

Fuel-Gathering Scoops

The British Interplanetary Society's Project Daedalus team considered the Jovian atmosphere to be the most promising source of ^{3}He for their fusion micropellet interstellar probe. But the "runner-up" approach suggested by the team was to point a Matloff/Fennelly electric-magnetic ion scoop toward the Sun and capture ^{3}He and deuterium from the solar wind. In this application, the scoop would orbit the Sun.

The solar-wind velocity is typically 300 km/sec and it has an ion density of 10^6 protons/m^3. For a solar wind deuterium/hydrogen ratio of 1.4×10^{-5} and a ^{3}He/hydrogen ratio of 3×10^{-5}, a negative 1000 coulomb charge on the scoop, and 10% efficient collection, about 0.01 kg/sec of ^{3}He and 0.003 kg/sec of deuterium would be obtained. Gathering enough fuel from the solar wind for a large fusion rocket would clearly take decades. Though deuterium is relatively easy to extract from the Earth's oceans, the ^{3}He isotope is much more difficult to obtain terrestrially.

The long-charged cable would eliminate electron drag, and the deuterium and ^{3}He ions would be decelerated and neutralized using electrons stripped from singly-ionized ^{4}He (a common ion species in the solar wind). The push away from the Sun that the decelerated fuel would impart to the scoop could be counteracted by a low-thrust propulsion system such as an ion engine or a solar sail. Captured fuel would be stored for later use in fusion rockets or pulsed fusion vehicles.

A version of this scoop could also accomplish inflight fueling of a fusion rocket from what might be thought of as the "induced interstellar wind." Consider a 2×10^4 ton fusion rocket equipped with a -20 coulomb scoop moving through the ICM at 0.01 c, after burning its original supply of fuel. After 60 years of cruising with its ion collector on, enough deuterium and ^{3}He would have been gathered to double the starship's speed, even though about 30% of the fuel would have to be expended to counter drag during fuel collection. Since the exhaust velocity of the Daedalus fusion engine is 0.03 c, inflight refueling would be limited to velocities lower than this.

Deceleration Sails

Although many investigators have denied the feasibility of ramscoops and thus cast doubt on ramjets and RAIR, no one seems to have dismissed their use as an aid in interstellar deceleration. Papers on electric or magnetic drag screens have been published by a number of authors, including Martin, Langton, Powell, and Roberts (10–13).

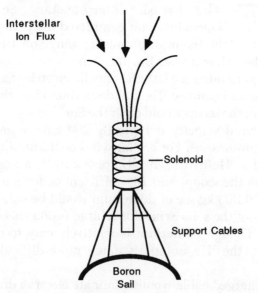

Figure 8.7 Deceleration by magnetic scoop.

Figure 8.7 presents a method of deceleration based on the Matloff/ Fennelly magnetic scoop. Interstellar ions collected by the scoop are deliberately bounced off a high melting-point (2600°K) boron sail towed aft of the scoop, thus dissipating the ship's kinetic energy. Several decades would be required to decelerate from 0.1 c to 0.002 c. The dynamics of sail-aided deceleration are presented in Technical Note 8–5.

Though the Matloff electric/magnetic scoop might serve the purpose, Figure 8.8 presents an adaptation of the Matloff/Fennelly scoop as a decelerator (8, 9). Sail charge is $+0.02$ coulomb and the cable charge is

Technical
Note
8–5

Magnetic Scoop Deceleration

Consider the magnetic scoop and boron-sail arrangement shown in Figure 8.7. Unaccelerated interstellar ions, after collection by the scoop, impact against the boron sail. Sail curvature reduces interactions between incoming and reflected particles.

The boron sail has a radius of 5 km, a thickness of 10^{-6} m, and a mass of 2×10^5 kg. To prevent the sail temperature from exceeding the 2600°K melting point of boron, a scoop field radius less than 1400 km is required at a ship velocity of 0.2 c, with the conservative assumption that all particle kinetic energy is converted into heating the sail.

From momentum conservation for elastic and inelastic collisions of ions with the sail, spacecraft deceleration is written:
elastic:

$$\dot{V}_s = 2A\rho m_p \frac{V_s^2}{M_s}$$

inelastic:

$$\dot{V}_s = A\rho m_p \frac{V_s^2}{M_s}$$

where A = scoop field area, ρ = interstellar ion density, m_p = proton mass, V_s = ship velocity, and M_s = ship mass.

If the effective area of the scoop is allowed to increase as V_s decreases and $M_s = 4 \times 10^6$ kg, deceleration from 0.2 c to 0.001 c for inelastic collisions requires 23.6 years. Since 0.001 c is still a hefty 300 km/sec, some additional terminal deceleration—perhaps with a solar sail—will still be necessary. During this time, the ship travels 0.52 light years. Because of the V_s^2 factor in the above equations, magnetic (and electric) scoop deceleration efficiency falls off rapidly as the ship decelerates.

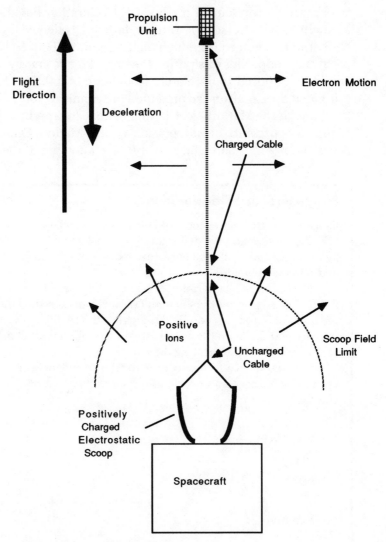

Figure 8.8 Charged deceleration sail.

minus 0.002 coulomb. The auxiliary propulsion unit is necessary to keep the cable from collapsing at decelerations greater than 1.5×10^{-4} g. The authors have in recent years assessed solar sails for interstellar flight, in the course of which author Mallove realized that a metallic solar sail could be deployed and electrically charged to play the role of the positively charged deceleration sail of Figure 8.8.

Other Ion Scoop Applications

At least two other applications of ion scoops *within* the Solar System are worth mentioning: (1) Richard Johnson's suggestion for shielding an orbiting space colony from cosmic rays (17); and (2) Mounting the Matloff/Fennelly magnetic scoop in a lunar crater to collect hydrogen from the solar wind. Combining the collected hydrogen with oxygen obtained from oxygen-rich lunar rock would provide water for lunar colonists and hydrogen for their chemically fueled Moon cars and commuter rockets.

9

Other Novel Advanced Propulsion Concepts

Who can guess what strange roads there may yet be on which we may travel to the stars?

Arthur C. Clarke, *The Promise of Space,* 1968

There is nothing so big nor so crazy that one out of a million technological societies may not feel itself driven to do, provided it is physically possible.

Freeman J. Dyson, 1965

The dream would not down and inside of two months I caught myself making notes of further suggestions. For even though I reasoned with myself that the thing was impossible, there was something inside me which simply would not stop working.

Robert H. Goddard, 1904 diary entry

It has been a long road from rocketry to ramscoops. We are now familiar with the basic concepts and design trade-offs of starflight propulsion systems that have received the most attention in the literature to date. Now it is time to dream a little bit more, not too wildly at first, however. Chapter 13 will deal with the really "far out" matters—ideas truly between fact and fancy. For the moment, consider just a few of the more novel interstellar flight concepts that have graced the pages of some very staid journals these past few decades.

Energy from Astronomical Bodies

Astronomical bodies have enormous kinetic energy as they whirl in their orbits. (Kinetic energy is energy of motion, where m = mass of the body, and V is its velocity: $E = \frac{1}{2} mV^2$.) Even mundane spheres such as the Earth and Moon possess extraordinary energy of motion. Finding ways to extract even a microscopic fraction of that energy and convert it to propulsive ends has been a favorite theme of investigators for some time. More often, the quest is referred to as an attempt to "extract gravitational energy" from astronomical systems. It is noteworthy that the first interstellar spacecraft did just that when they were "gravity whipped" by some of the outer planets of the Solar System and flung into interstellar space. This is ironic because gravity is an incredibly weak force, about 40 orders of magnitude weaker than the electrostatic force between two particles. But when mass gathers together in astronomical bodies, its collective attractive force can be staggering, and it permits the fantastic storage of energy (even by starship standards!) in the orbital motion of celestial bodies.

Stanislaw Ulam, of nuclear pulse propulsion fame (see Chapter 4), wrote in the late 1950s and early 1960s on the possibility of space vehicles extracting energy from astronomical bodies (1,2). His approach was abstract and theoretical, and he asked whether in an idealized three-body system (one of the bodies being a rocket of small mass) could one "obtain a velocity arbitrarily large—that is, close to the velocity of light?" He concluded that though this might be possible in theory, the slowness of maneuvers to build up to such speeds in a multibody system would make the process impractical.

The next person to consider the problem was the ever-inventive Freeman Dyson, whose paper on "Gravitational Machines" appeared in 1963 in the first major technical book on SETI, a field that was then called "interstellar communication"(3). Dyson has been a pioneer in considering what extraterrestrial civilizations might do in the way of "astronomical engineering," and—by implication—what our civilization might aspire to (4). As an example, Dyson imagined a double star with each of the components having a mass equal to that of the Sun and revolving with velocity, V, about a common center in a circular orbit of radius, R (Figure 9.1).

Dyson imagined a craft dispatched with a small velocity onto the trajectory indicated in the figure. Closely approaching the star traveling opposite to the direction of the approaching craft, the ship would be whipped around the star and flung outward to arrive with velocity 2V, not at the departure point, but roughly as far away from the double star. The final velocity of the craft and the energy that could be extracted from the double star system are given in Table 9-1 for various hypothetical stars and conditions. The figures in the Table are based on Dyson's

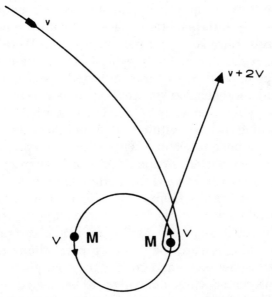

Figure 9.1   Dyson's gravity machine. (Courtesy Robert L. Forward)

formulae of Technical Note 9–1, and, for simplicity, assuming each star of a pair to have identical mass.

It is amazing that Dyson put forth these speculations about five years before the existence of then hypothetical neutron stars had been confirmed by radio astronomers. Neutron binary stars would permit extraordinary starship velocities if a way could be found to nullify the extreme tidal forces that would normally tear apart objects approaching them closely. It is not out of the question that someday we or an extraterrestrial civilization might engineer a Dyson gravitational machine for starflight or other purposes. Perhaps *they* already have! It is easy to imagine that a number of double star systems located in close proximity could be used as a multistage "gravitational accelerator" (our term) to reach the high velocities about which Ulam speculated.

Table 9-1   Dyson's Gravitational Machine

Pair of Stars	Departure Velocity
White dwarf stars, each with a diameter of 20,000 km, one solar mass, and with a combined orbital period of 100 sec.	0.009 c
Neutron stars, each with a diameter of 20 km, one solar mass, and with a combined orbital period of 0.005 sec.	0.27 c

We space travel neophytes are already using gravitational energy to hurl once merely interplanetary spacecraft into the interstellar abyss. The late Kraft Ehricke, a former member of Wernher von Braun's Peenemunde rocket team, wrote extensively on the use of multiplanet gravity whips to send probes into interstellar space at speeds that he claimed could approach 200 km/sec (5,6). In our research, we previously had derived results that were in agreement with Ehricke's conclusion that approaching the Sun within 0.01 AU of its center (about two solar radii) on a hyperbolic trajectory (with *no* powered perihelion maneuver) would yield a Solar System escape velocity of 180 km/sec (7).

Unfortunately, recent work by author Matloff and Kelly Parks has shown that these results were optimistic by almost a factor of two, because of a subtle error in the manner of computing rebound velocities (8). Using corrected equations, Matloff and Parks computed the result of a worldship falling toward the Sun on a parabolic trajectory—perhaps from the Oort comet belt—and grazing the solar photosphere (0.005 AU from the Sun's center) while applying a 10 km/sec powered maneuver at perihelion. The hellish penetration of the solar atmosphere at 600 km/sec would occur in a relatively short but sizzling half-hour period. With the 10 km/sec maneuver, the starship would exit the Solar System with a hyperbolic excess velocity of 110 km/sec, and with a 20 km/sec perihelion maneuver, V_∞ would move up to 156 km/sec, that is, 8 to 12 millennia to Alpha Centauri.

<table>
<tr><td>Technical
Note
9–1</td><td>

Formulas for Dyson Gravitational Machines

Consider two equal mass components of a double star system, revolving about a common center in a circular orbit of radius, R. The orbital velocity, V, of the stars is:

$$V = \sqrt{\frac{GM}{4R}}$$

The departure velocity of the starship, $V_\infty = 2V$.

Dyson's formula for the total energy extractable from the gravitational field when the stars have spiraled in toward one another and are separated by $4r_s$, where r_s is the radius of each star:

$$E_g = \frac{GM^2}{8r_s}$$
</td></tr>
</table>

Electromagnetic Launchers

Arthur C. Clarke introduced "electromagnetic launching" to the world in 1950, analyzing it in his prescient article, "Electromagnetic Launching as a Major Contribution to Spaceflight" (9). Clarke acknowledged that others had considered the concept in the past, but he was surely the person who made the idea tangible for the first time. He proposed a system that would accelerate payloads from a kind of electromagnetic catapult on the surface of the Moon. When introduced, Clarke's idea was, of course, far ahead of its time, but now in an era of so-called "mass drivers" favored by advocates of space colonization or of "kinetic kill" weapons designed for orbital defense against ballistic missiles, electromagnetic launching may have come of age. Mass drivers and kinetic energy weapons are merely special limited applications of electromagnetic launching. Free of these "mundane" pursuits one or more generations hence, electromagnetic launching may propel starships.

In his 1950 paper, Clarke described the mechanics of a three-kilometer-long electromagnetic launcher on the Moon which in several seconds would accelerate payloads at 100 g's to lunar escape velocity (2.3 km/sec). Then he wrote of electromagnetic launchers deployed in free-space, remarking, "Since there seems no physical limit either to the lengths or accelerations which might be utilized in this case (if sufficient power were available) such projectors may conceivably play a part in the development of interstellar flight."

Despite practical limitations imposed by current technology, others have also noted the enormous potential of electromagnetic launching for advanced space propulsion, at least for dispatching small robotic probes. They have recognized that electromagnetic launching shares with beamed power propulsion a key attribute: propellant need not be accelerated, only the payload. Electrical energy input to the electromagnetic catapult could be provided by storage reservoirs consisting of massive flywheels or other kinds of energy "batteries," such as electrical capacitors.

Winterberg has tackled the problem of propelling superconducting solenoids with a traveling magnetic wave accelerator (see Figure 9.2) (10,11). His main objective was to demonstrate the feasibility of reaching speeds of 100 to 1000 km/sec along path lengths of a few kilometers, the speed range required to create a laboratory-scale controlled thermonuclear explosion when the projectile collided with a fusion fuel target. One of the central difficulties with some kinds of "railgun" accelerators had been the high-temperature radiating gas formed when the projectile was in physical contact with the accelerator. Winterberg

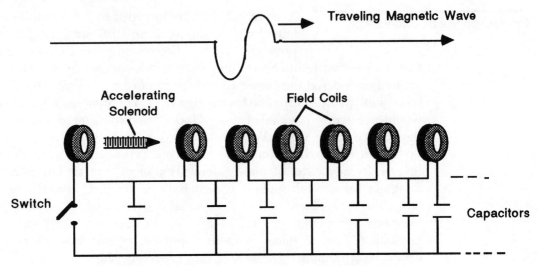

Figure 9.2 Traveling wave electromagnetic accelerator.

proposed to avoid contact of the accelerating magnetic dipole with the launcher by appropriate electromagnetic deflectors and control systems. Apparently such a feat is theoretically possible with present technology, if, in fact, it has not already been carried out.

Lemke later directly addressed the application of electromagnetic launching to interstellar flight, opting for a superconducting dipole accelerator of literally astronomical proportions, 10^8 km long or two-thirds of an AU (12). For ten days between successive launchings of 10-ton payloads, a huge solar collecting array (200 km × 200 km) would store the requisite energy to boost each payload to 0.33 c with an acceleration of 5000 g's. Such large linear structures should be built beyond the orbits of Pluto and Neptune to avoid the bending moments caused by gradients in the Sun's gravitational field. But to rely on solar power, the accelerator would best be sited in the inner Solar System. Lemke proposed that numerous probes containing telescopes be dispatched toward neighboring stars—shotgun-style—with the expectation that some probes would pass close enough to extrasolar planets to return useful data.

Unquestionably, electromagnetic launching is still in its infancy, and there are many unresolved matters of power supply, power phasing, and guidance of probes down an accelerator path. But because of the inherent advantages of the concept, it seems an infant well worth nurturing.

Pellet Stream Propulsion

In 1979, Clifford E. Singer of Princeton University's Plasma Physics Laboratory proposed a refreshing interstellar propúlsion concept that makes use of electromagnetic launching, but that also has some of the attributes of beamed power and nuclear pulse propulsion (12). He did not suggest the electromagnetic acceleration of complete payloads, but of streams of small pellets that would impact a starship and transfer momentum to it through a variety of possible mechanisms. On arriving at the starship, pellets might be scattered rearward elastically (or simply stopped) by means of powerful magnetic fields. Alternately, the high-velocity pellets might disintegrate on impact with the target and be transformed into a plasma that would be exhausted rearward—as in a nuclear pulse vehicle like Daedalus (see Chapter 4).

Singer's pellets would be in the mass range of 3 to 100 grams. Typically, these superconducting pellets would be aimed at their starship target during a significant fraction of the mission, though there would also be a coasting phase. Singer's performance analysis envisioned an accelerator 10^5 km long deployed in interplanetary space, one that would produce a constant pellet acceleration of 0.3 to 4 "megagravities" (Mgrav = 10^6 g's). He noted that such accelerations had already been achieved in the laboratory with a "rail gun" accelerator boosting one-gram pellets over a four-meter path. Singer speculated about a pellet-stream mission comparable to the Daedalus flight to Barnard's Star (5.9 ly), that is, a fly-through probe velocity of 0.12 c and a 50-year flight time with 450 tons of payload. The required power source would have to average 15,000 gigawatts over a 3-year period to launch two 2.8 gram pellets each second at 0.25 c.

A major consideration in this propulsion system is accurate collimation of the pellet stream. Singer claims this as one of the system's great virtues. Not only could accurate collimation be achieved, he says, with several dozen initial correction stations spaced 340 AU apart, but also the starship itself could adjust its position to remain in the stream. For example, the starship might employ radar to detect approaching pellets. The measurement stations—either predeployed or "shed" by the starship—would measure by optical means the location of each pellet passing through it and relay commands to more distant stations to correct the flight path with electrostatic or magnetic fields. Singer contends, "Only the *relative* velocity dispersion between one pellet and the next makes a significant demand on the [starship auxiliary] propulsion system."

In his first pellet-stream paper, and in his second one in response to critics, Singer dealt with alleged problems stemming from the interstellar medium (13,14). He concluded that pellet-stream dispersion due to the impact of interstellar grains was not unreasonable for 1 to 1000 gram pellets, though it could be a problem for lighter ones. Likewise, he concluded that pellet charging and interaction with the galactic magnetic field would not impose a fundamental barrier to the concept, much less would any conceivable gravitational influences or differential illumination by starlight.

Workable concept or not, the advent of the pellet-stream propulsion idea several decades after the beginning of serious starship speculation illustrates again how easy it is to overlook "obvious" interstellar flight concepts. What other propulsion gems may be waiting to be found, buried in the armamentarium of twentieth-century technology!

Brainstorms

Not all ideas for starflight propulsion systems have been developed even to the extent of the latter few proposals. Witness the fruits of a few odd brainstorms:

Elastic Collision Propulsion (Mallove, circa 1973)

If a massive body, either astronomical or artificial, is made to collide *elastically* (involving little or no energy loss) with a body of smaller mass, a significant velocity increment will be imparted to the less massive object. In fact, the limiting velocity increment is two times the relative velocity of the more massive object (see Technical Note 9–2). Thus, the concept bears some resemblance to Dyson's gravity machine idea. The momentum transfer might be mediated by electrostatic, magnetic, or even mechanical spring interactions, a prime design consideration being the minimization of energy losses by irreversible effects.

Now if a series of bodies of successively smaller mass are arranged to undergo a "chain reaction" of collisions, it is possible, in theory, to build up a significant velocity in the least massive body. This would be a kind of "space billiards." As an example of what we might ultimately achieve with the concept, imagine that we have commandeered an asteroid with twice the mass of Icarus ($2 \times 5 \times 10^9$ tons). If the 10^{10}-ton body collided elastically with a 10^8-ton astronomical object, which in turn collided with a 10^6-ton mass, that with a 10^4-ton object, that with a 10^2-ton manmade body, and finally with a 10-ton body, the last mass would have a velocity 32 times the asteroid velocity relative to the first body. If the

Technical
Note
9-2

Elastic Collision Propulsion

Consider the case of a large mass, M, with velocity V_o making a perfectly elastic collision with small mass, m, initially at rest. After collision, the large mass has velocity, V, and the small mass has velocity, u. For a perfectly elastic collision, the momentum and energy conservation equations lead to an expression for u:

$$u = \frac{2V_o}{1 + \dfrac{m}{M}}$$

The limiting velocity, u, as m/M approaches zero is, of course, $2V_o$.

Extending this single elastic collision to multiple collisions, we find that the velocity of the nth mass after collision with the $(n-1)^{th}$ mass is simply:

$$V = 2^{(n-1)}V_o$$

asteroid velocity is 35 km/sec, our 10-ton payload departs the Solar System at 1000 km/sec or about 0.003 c. It is, of course, far easier said than done to make these successive collisions nearly elastic or to set up the cascade of objects in the first place.

"Scissors" Propulsion (Mallove, circa 1976)

The geometric intersection point of the collapsing blades of a pair of scissors could, in theory, exceed the velocity of light if the blades were sufficiently long. If we were to arrange the equivalent of two very large "blades" in space with a differential angular velocity about their pivot point, the region of the *geometric* intersection of their edges could attain large velocities, even though the absolute magnitude of angular velocity of the blades was not structurally prohibitive. Perhaps a pair of such massive scissor blades could apply forces to a suitably arranged probe—electromagnetically suspended near the vertex of the scissors—and accelerate it to high velocity (Figure 9.3) (17).

The giant scissor arms might be fashioned from a reformed asteroid, which could launch multiple probes as kinetic energy was extracted from the rotating body. Of course, the scissor arms would have to be spun up by other propulsive means, but the spinning could be accomplished over a long period of time, for example, by electric thrusters at the arm tips,

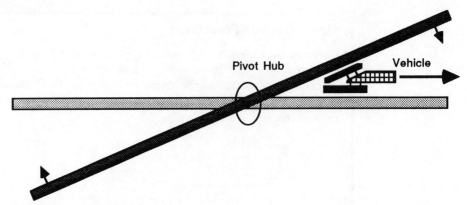

Figure 9.3 The "scissors" launcher.

propellant being supplied through the arms from a nonrotating tank near the pivot. The probe would consist of payload and superconducting magnets that would suspend a probe in the vertex of the scissors as it rode down an electromagnetic guideway. In effect, the probe would be "flicked" into space like a cherry pit pressed between two fingers.

These admittedly wild ideas support the notion that starship propulsion concepts often hide where one least expects them. When it comes to starflight, one is sold by experience on the virtues of brainstorming. One day, a starflight pioneer gazing at his or her surroundings may fashion a now undreamt highway to the stars, a "spaceship of the mind" that will become reality. As we "thrust home," we offer this oration from Edmond Rostand's play, *Cyrano de Bergerac*, as encouragement for the brainstorms of our readers (18):

You wish to know by what mysterious means
    I reached the moon?—well, confidentially—
    it was a new invention of my own.

    . . .

    I imitated no one. I myself
    Discovered not one scheme merely, but six—
    Six ways to violate the virgin sky!

    . . .

    As for instance—Having stripped myself
    Bare as a wax candle, adorn my form
    With crystal vials filled with morning dew,
    And so be drawn aloft, as the sun rises
    Drinking the mist of dawn!

. . .

Or, sealing up the air in a cedar chest,
Rarefy it by means of mirrors, placed
In an icosahedron.

. . .

Again,
I might construct a rocket, in the form
Of a huge locust, driven by impulses
Of villainous saltpetre from the rear,
Upwards by leaps and bounds.

. . .

Three,
Smoke having a natural tendency to rise,
Blow in a globe to raise me.

. . .

Four!
Or since Diana, as old fables tell,
Draws forth to fill her crescent horn, the marrow
Of bulls and goats—to anoint myself therewith.

. . .

Five!
Finally—seated on an iron plate,
To hurl a magnet in the air—the iron
Follows—I catch the magnet—throw again—
And so proceed indefinitely.

. . .

The ocean!
What hour its rising tide seeks the full moon,
I laid me on the stand, fresh from the spray,
My head fronting the moonbeams, since the hair
Retains moisture—and so I slowly rose
As upon angel's wings, effortlessly . . .

Cyrano in Cyrano de Bergerac, Act III

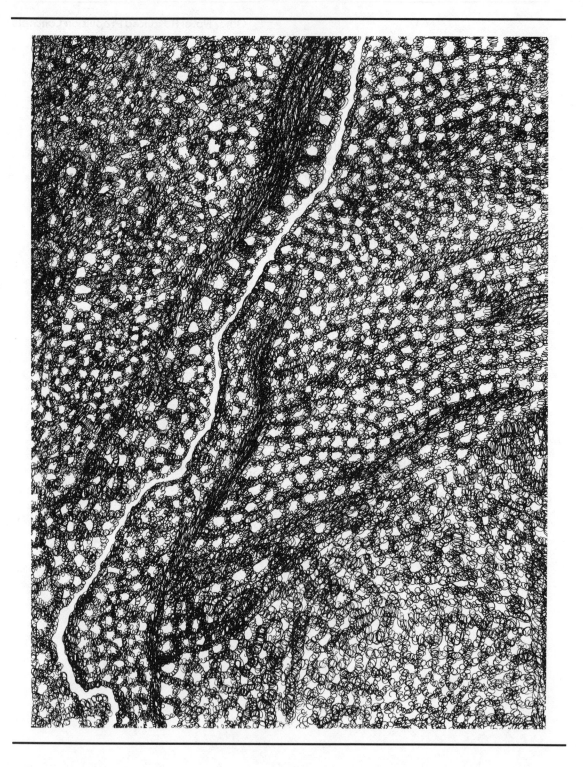

10 Interstellar Trajectories

While he was suspended there, a fantasy took shape in the mirrors of his mind, an image sharp enough to shut out the surrounding scene. A mechanical device materialized from nowhere, functioning perfectly. Faster and faster it whirled until it began to lift, twirling and spinning above Worcester and sickness and spaniels and fruit trees, upward into space.

Milton Lehman, *This High Man: The Life of Robert H. Goddard*

Point and Shoot, or . . .

The road to the stars may be paved with good intentions, but they will not get a starship there that wastes its precious energy with ill-timed maneuvers. A starship's thrusting or other kind of propulsive energy expenditure must, above all, be appropriate to its particular kind of propulsion system. Now at first glance, nothing would seem simpler than to optimize an interstellar trajectory: simply point the trusty starship in the direction of the target; ignite the engine, fire up the beam, or unfurl the sail; and say "good-bye" to the Solar System, perhaps forever.

After accelerating and wringing out as much velocity as possible, the engine (if any) is perhaps turned off, and a cruise phase begun. Approaching the destination, the decelerator system is engaged to brake the ship to the appropriate interplanetary orbital speed. Alternatively, with a propulsion system sufficiently capable, the ship might accelerate for half the flight and decelerate for the remaining half, thus dispensing altogether with the cruise phase.

The reason for this apparent simplicity is the basic difference between interplanetary and interstellar flight. Interplanetary space is dominated by the gravitational fields of a sun and planets; interstellar space is essentially free of significant perturbing gravitational influence, and what slight gravitational deviation there is on a trajectory is made moot by the dominance of the high-energy, nearly straight flight path. In low-speed interplanetary flight, we have grown accustomed to the complex tailoring required to design a Keplerian trajectory linking two planets, which themselves move on sluggish elliptical paths. But a fast starship free of the bonds of the Sun and its planets should be able to zip off on a straight line to the stars, right?

Wrong! The design of optimal interstellar trajectories is actually more demanding than pointing and shooting because of limitations imposed by stellar motions, technology, economics, and the ship's payload. It is also necessary to determine precisely what it is about the trajectory that should be optimal. Because optimizing interstellar trajectories is now more a paperwork art than a rigorous science, the considerations that go into optimizing the flight paths of various types of starships are merely described.

Gravity-Assist Trajectories

The only interstellar missions so far have been the Pioneer 10 and 11 and Voyager 1 and 2 robot explorers launched by the United States during the 1970s. All of these probes relied on the gravity-whip assistance of the Solar System's giant outer planets to achieve very modest interstellar cruise velocities. At best, these are marginal probes of transstellar space, for tens of thousands of years will elapse before they cross the 4.3 ly gulf between the Sun and Proxima Centauri. (Of course, the probes are not actually targeted on that star, but the distance is a convenient gauge.)

The dynamics of planetary gravity-whip assist trajectories were mentioned in the previous chapter, and in Chapter 6 the more fruitful solar-gravity assist trajectory was discussed, in which a large velocity is imparted to a spacecraft during a close approach to the Sun. If the objective is to optimize the interstellar velocity of a Pioneer or Voyager-class probe, it is necessary to take advantage of the fortuitous alignments of Jupiter and Saturn that occur at intervals of about 20 years. Every 180 years or so, Uranus and Neptune line up in the correct order with Jupiter and Saturn, allowing an even greater velocity boost to a departing probe.

This is the so-called "Grand Tour Alignment," which will allow Voyager 2, launched in 1977, to have visited four planets—Jupiter, Saturn, Uranus, and Neptune—by the time it exits the Solar System. It will be the middle of the twenty-second century before this mission could be repeated. But with the much more advanced propulsion systems available then, we would expect only a symbolic rerun of that ancient mission—like the reconstructed Mayflower crossing the Atlantic (1). If one had to control the direction of the exit trajectory as well as speed, that is, aim at a particular star, a long wait until an appropriate grand tour alignment occurred would be necessary. Depending on the location of the target star, its motion, and the desired cruise speed, the wait might be thousands of years (see Technical Note 10–1).

Of course, no effort was made to aim the Pioneers and Voyagers toward specific stars. According to a comprehensive study by NASA researchers at JPL in 1984, Pioneer 10 is exiting the Solar System north of the *ecliptic* (the plane in which Earth orbits) near the constellation Auriga (2). Pioneer 11 also exits north of the ecliptic between Ophiuchus and Capricornus. Voyager 1 goes farther north, moving from Boötes toward Ophiuchus. Its sister ship Voyager 2, traveling near the ecliptic plane before its final planet encounter, will pass Neptune in August 1989 and depart southward toward the constellation Tuscana. All four probes are moving at roughly 2 to 4 AU per year, and at this rate the fastest among them—Voyager 1—would reach Alpha Centauri (if it were aimed at it) in about 60,000 years. But as we have seen, the stars are not stationary on time scales of thousands of years. In tens of thousands of years, the shift in the apparent positions of the nearby stars will be dramatic, caused by the Sun and nearby stars moving at differing relative velocities around the hub of the Milky Way galaxy.

Although staying in radio contact with these craft is unlikely much beyond the year 2015, when Pioneers 10 and 11 and Voyagers 1 and 2 will be respectively 110, 90, 130, and 110 AU from the Sun, the JPL team

Technical Note 10–1

Planetary Alignments

To get an intuitive feel for planetary alignments, consider the case of Jupiter and Saturn. Jupiter takes 11.86 (Earth) years and Saturn requires 29.46 years to circle the Sun. Approximating the orbital periods of the two planets as 12 and 30 years, we find that Jupiter and Saturn move through about 30 and 12 degrees per year respectively.

Discounting the inclination of the planet orbits to the ecliptic (1.3° for Jupiter and 2.5° for Saturn) and their non-circular elliptical orbits, assume that we first observe Jupiter and Saturn when they are on the same radial line from the Sun. After T years, the two planets will line up once again. T is found approximately from the relation:

$$30T - 360 = 12T$$

Hence, T is about 20 years. During 20 years, Jupiter turns through about 600° (360° + 240°) and Saturn through 240°. The planets will be aligned once more, but 240° (⅔ of the angular distance around the ecliptic) from their original positions.

surveyed stellar motions tabulated in the Gliese Catalogue and the General Catalogue of Stellar Radial Velocities (3,4). They computed close encounters of these spacecraft with stars for the next million years, even though present observational errors in stellar positions and motions will dynamically propagate into more pronounced uncertainty as time goes on.

The closest known stellar encounter by Pioneer 10 occurs in 33,000 years, when it will pass within 3.3 ly of Ross 248, a red dwarf star. Pioneer 11 and Voyager 1 will each pass about 1.65 ly from another red dwarf, AC + 79 3888, in about 40,000 years. In 39,600 years, Voyager 2 will be within 1.25 ly of Ross 248, in 47,000 years within 2.75 ly of AC + 79 3888, and in 358,000 years, Voyager 2 will come within 0.8 ly of Sirius, presently the brightest star in the skies of Earth.

Like "spaceship Earth" the Sun too is a spaceship, *literally* a starship as it moves among neighboring stars! Let it be recorded that this is where the Fulleresque phrase, "Starship Sun," was born. Considering the Sun as a spacecraft, the JPL researchers discovered six known stars that will pass within 3 ly of Sol during the next million years. Three of these will come within 1.5 to 1.7 ly, and about the year 815,000, one (DM + 61366) will be only 0.29 ly away! Although errors in stellar motion and position are large for this star, the K5 spectral class dwarf's *predicted* close approach to the Solar System might gravitationally disrupt the orbits of some comets in the Oort comet belt. Recall that physicists Luis Alvarez, Richard Muller, and others have suggested that such a disturbance might send swarms of comets toward the inner Solar System, causing impacts on the Earth and future mass extinctions of terrestrial species (5). (Of course, by then we'll be smart enough to save at least our own precious hides, or will we?) Because DM + 61366 is a Sunlike single star, it might well have a planetary system, and in a mere 800,000 years we and the possible inhabitants of the DM + 61366 planet system (if such exists) will have only a short interstellar hop for mutual visits.

Freeman Dyson suggested that gravity assist trajectories of another type may be useful to galactic civilizations (6). Suitably located cultures might use "gravity machines" consisting of double white dwarf stars for interstellar transport. Such a "gravity machine," as illustrated in Figure 9.1, could accelerate a starship to 1500 km/sec or more. An extremely advanced civilization might relay huge masses across the galaxy by creating such machines from otherwise "unused" binary stars or naturally occurring double stars. To that end, they might even perform advanced astroengineering to speed up stellar evolution and create conveniently placed co-orbiting white dwarfs.

Solar Sail Trajectories

As outlined in Chapter 6, an interstellar solar sail mission would have several phases. First, using electric propulsion (and possibly giant planet gravity whip assists) the ship would drive toward the Sun from its starting position in the outer Solar System. If solar-electric propulsion were used in this initial phase, the preperihelion trajectory would be energy-limited. Toward the end of the electric propulsion phase, the trajectory would be limited by the acceleration capability of the propulsion system because in the inner Solar System sunlight is much more intense than in the outer reaches.

During the close solar pass, acceleration limitations would be imposed by the tolerance of the payload, support cable strength, and sail thickness. The authors have published a computerized technique of trajectory optimization during this phase (7). After the perihelion pass, electric propulsion might be used to correct trajectory errors and to provide additional acceleration. Because solar energy falls off rapidly postperihelion, energy beaming might be required to counter the rapidly developing energy limitation.

Alternatively, postperihelion acceleration could come from a variant of Clifford Singer's pellet stream propulsion (see Chapter 9) (8). A solar-powered device near the Sun would launch a high-velocity stream of small particles (pellets) toward the departing starship. Acceleration would be optimized in two ways: (1) tight collimation of the pellet stream to insure collisions between the fast particles and the slower starship; and (2) the design of pellets and a starship shock absorber such that kinetic energy transfer is maximized and heat production minimized during the collisions.

Fusion-Rocket Trajectories

Fusion rockets have been the most thoroughly analyzed of interstellar propulsion systems, so it is not a surprise that more optimization studies have been done for fusion-rocket trajectories than for any other method of starflight. Suppose we choose to vary propulsion parameters to minimize the cruise time. Or perhaps we decide to accept a longer cruise and instead minimize the requirement for fusion fuel, thereby reducing mission cost. By lowering the engine exhaust velocity, we could perhaps reduce technological problems, but at the cost of increased mission time. These are examples of issues involved in trajectory optimization.

Technical
Note
10-2

The Calculus of Variations

The basic intent of the calculus of variations is summarized in the accompanying figure. This is the simplest, but by no means not the only kind of problem that can be treated with the calculus of variations (21).
Define the integral:

$$I = \int_{x_1}^{x_2} F(x,y,\frac{dy}{dx})dx = \int_{x_1}^{x_2} F(x,y,y')dx$$

for a function, y, defined between two fixed endpoints x_1 and x_2. The objective is to find a function, y(x), that will minimize or maximize the value of this integral. The first derivative of y with respect to x, dy/dx, is symbolized by y'. For what value of y(x) is the function a maximum or a minimum?

In the figure, $\overline{y(x)}$ is a path that is slightly varied from y(x), but with the constraint of having the same endpoints. After some manipulation, the maximum or minimum value of integral, I , can be shown to occur when the so-called *Euler equation* is satisfied:

$$\frac{d}{dx}\left(\frac{\partial F}{\partial y'}\right) - \frac{\partial F}{\partial y} = 0$$

Solutions of this equation are called *extremals* of the problem.

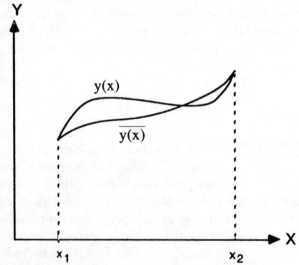

Basic principle of the calculus of variations.

The fusion-rocket trajectory designer would perform a number of engineering trade-offs to take these and other factors into account to arrive at an approach close to optimum. The mathematical technique called the *calculus of variations* is used to carry out these kinds of optimizations. Described at an elementary level in Technical Note 10–2, the calculus of variations works by assuming the trajectory endpoints or end conditions (start and finish) to be fixed. A number of parameters are then varied mathematically to arrive at an optimum path between these endpoints.

Pioneers in optimizing fusion-rocket trajectories include G. M. Anderson and Conley Powell (9–14). Although their work specifically addressed fusion rocketry, their analyses could, in fact, be applied to fission and antimatter rockets and even high performance nuclear-electric propulsion. Anderson, then a professor at the USAF Institute of Technology, published his optimization analyses in several papers between 1968 and 1974 (9–11). Although he focused on fusion rockets, he also extended his calculations to antimatter, fission, and electric propulsion.

Using the calculus of variations, Anderson derived relativistically correct results for a flight time limited to 40 years *and* a mass ratio limited to 10,000. He discovered that of all the rockets considered—antimatter photon rockets ($V_e = c$), ideal nuclear-fusion rockets ($V_e = 0.0893$ c), ideal nuclear-fission rockets ($V_e = 0.0388$ c), and conservative ion rockets ($V_e = 0.000316$ c)—only the ion rocket is inherently not capable of carrying out the mission.

Anderson's study assumed zero initial and final velocities and minimum-time trajectories. He defined a critical parameter, the "mode switching mass" (MSM), at which the thrust program changed from a thrust-limited to an acceleration-limited mode. For each rocket type considered, except the ion rocket, Anderson discovered the maximum flight distances for which the constraints of flight time and mass ratio are simultaneously satisfied. He determined the associated MSM for each rocket type.

The hypothetical (and perhaps impossible) photon rocket is, of course, fastest and has the longest range. During a 40-year flight (from a crew member's point of view) the antimatter ship could achieve a speed of 0.9998 c and traverse 1200 ly. The fusion ship would traverse 15.3 ly at a peak velocity of 0.3895 c, and the fission rocket would reach 0.1766 c while traveling 6.5 ly.

Conley Powell, while at the University of Kentucky and the University of Tennessee, expanded on Anderson's pioneering efforts (12,13). He found that fusion drives are acceleration-limited rather than energy-limited for flights to nearby stars. Fusion rockets will therefore most likely be nonrelativistic. In one optimization, he allowed the rocket exhaust velocity to be time-varying over a wide range to maximize cruise velocity. The ship's fuel-burnout velocity was only slightly greater than if the exhaust velocity had been held constant.

Powell later considered minimizing flight time for a multistage flyby of Barnard's Star (13). He constrained each stage to have the same engine mass/power ratio, payload fraction, exhaust velocity, and initial acceleration. To achieve a 40-year flight time, an exhaust velocity of 8000 km/sec and a mass/power of 0.5 kg/megawatt are required! Nuclear pulse propulsion—of the Orion or Daedalus type—should be considered for such optimized missions.

Powell performed another optimization study with Rajendra Prasad Mikkilineni, considering a single-stage starship with a fixed mass ratio and a constant engine power expended over a fixed distance (14). Their objective was to determine the exhaust velocity program that minimizes flight time in the ship's frame of reference. At least for high nonrelativistic velocities, the reduction in flight time due to exhaust velocity programming was shown to be a function only of the given mass ratio.

Giovanni Vulpetti of Telespazio in Rome has also considered "nonrectilinear" relativistic trajectories in which payload-splitting and high-energy midcourse maneuvers would be used to allow fly-by probes on a common "bus," for visits to two stars per mission (18–20). Midcourse optimal trajectories were considered for paired fly-bys of Barnard's Star–61 Cygni, Alpha Centauri–Barnard's Star, Epsilon Eridani–61 Cygni, and Barnard's Star–Epsilon Eridani. Vulpetti has also considered time and energy optimization and optimization with multiple propulsion techniques.

Ramjet Trajectories

In Poul Anderson's novel *Tau Zero*, the fusion ramjet starship *Leonora Christine* is inadvertently committed to a circum-universe flight when it loses its decelerator during an encounter with a small, uncharted dust-rich interstellar nebula (a "nebulina") (15). But as Robert Bussard and Carl Sagan have suggested, an actual ramjet might deliberately target certain types of nebulae to maximize velocity and minimize flight time (16,17). The nebula most sought by a ramjet commander would be a dense ionized hydrogen (HII) region. A ramjet committed to a 1000 ly-transgalactic flight might well be able to afford detours of tens of light years to seek out such nebulae near the line of flight. A ramjet derivative, the electric or magnetic drag screen for interstellar deceleration, would perform optimally as the ship began to encounter the stellar wind of the destination star. Happily, this is just when deceleration is most needed.

11

The Interstellar Medium

O God! I could be bounded in a nutshell, and count myself a king of infinite space, were it not that I have bad dreams.

William Shakespeare, *Hamlet,* Act II, Scene 2

Dreaming about starflight has at least one occupational hazard, apart from the obsessive questing that goes with the territory: bad dreams, very bad dreams. The magnificent high-performance starship has been built on paper and is now ready for "cutting metal." Suddenly someone calculates anew that the interstellar medium is certain to erode, cook, or otherwise destroy our high-speed craft. More often than not, this sniping comes from those whose hobby is trying to prove that starflight will forever be impractical. This is for the greater good, however, because it serves to keep the starflight community honest with itself.

There is no doubt that this bad dream has some basis in fact because all starships will have a significant interaction with the tenuous broth of molecules and other particles between the stars. The higher the speed, the worse the problem. To design a starship without thorough knowledge of the interstellar medium would be almost as foolish as trying to build an aircraft without considering the properties of Earth's atmosphere or planning an ocean-going vessel without thinking about the properties of sea water. One should not pretend—as some have—that vacuum alone reigns supreme in interstellar space.

Ways that starships may be fueled or decelerated by the interstellar medium have already been reviewed. At quasi-relativistic velocities in particular, erosion by rare interstellar dust grains may become a significant problem. Another possible limitation on starship velocity is the induced flux of cosmic radiation as an inhabited starship smashes through the interstellar hydrogen. But before considering the engineering aspects of starship-medium interactions, a review of what is known about the interstellar ocean is necessary: its gas, dust, cosmic rays, and magnetic fields, and who knows what else!

Properties of the Interstellar Medium

Although the novice sky watcher may not have thought of it in this light, a beautiful nebula visible to a Northern Hemisphere observer with a small telescope or a good pair of binoculars is, in fact, an interstellar gas cloud. Viewed on a winter's evening, Messier 42, or simply M42, is located just below the hunter's belt in the constellation of Orion. At low magnification, M42 is a beautiful blue-green cloud, lit by young, hot stars within it. The nebula is a star nursery, one of the youngest such birthing grounds in the sky. New stars are born from gravitational and other instabilities within it, much as the Sun condensed in a similar nebula about five billion years ago. M42 is a youngster itself, being no older perhaps than several tens of thousands of years.

Observational and theoretical studies of the interstellar medium have not been carried out, of course, to design starships, though perhaps this will become another reason for the science before too long. The purpose of many of these studies has been to learn how to correct measurements of starlight for the effects of *interstellar extinction*—the *absorption* and *scattering* of starlight along its path to us.

When a beam of light travels through any medium—be it air, water, or the gas and dust between the stars—several processes occur that weaken the intensity of the transmitted light. Collectively these are called extinction or attenuation and include the separate effects, scattering and absorption. Scattering is due to the interaction of light photons with atoms, molecules, and larger particles. When a light beam interacts with a particle, part of the beam's direction may change. Molecules in Earth's atmosphere scatter blue light more strongly than red light, thus explaining the youthful question (for which few adults have the answer), "Why is the sky blue?" (Technical Note 11–1)

In absorption, photons are "stopped" when they reach a suitable atom or molecule. The energy of the photon is taken up by the atom or molecule as an excited electron, vibrational, or rotational state. Absorption results in a dark line imposed on the multicolored spectrum of white light, a line corresponding to the wavelength (color) of the absorbed photon. Later, the atom or molecule may emit one or more photons as it returns to its ground or unexcited state.

Interstellar scattering between wavelength (λ) 0.3 and 1 micron (10^{-6} m) varies roughly inversely with the wavelength, rather than as $1/\lambda^4$, which is the case for Rayleigh molecular or ultrafine particle scattering. The main scattering agents in the interstellar medium are therefore not molecules and atoms, but particles. Particles 0.1 micron in size would result in a scattering variation: $1/\lambda$. These are the interstellar dust grains.

Because blue light of shorter wavelength is scattered more than red light, a star across the galaxy will seem redder than its closer twin. This is called *interstellar reddening*—a simple observation that proves the existence of the interstellar medium. Others are the presence of bright and dark nebulae, the lack of observed distant galaxies in the dust-laden plane of the Milky Way, spectral lines of interstellar origin, *polarization* of starlight, and radio emission from interstellar gas.

The explanation of bright and dark nebulae owes much to the pioneering work of Bengt Stromgren in the 1930s (3). A dark nebula is a cloud of neutral hydrogen gas with a density about 10^9 atoms/m^3 and a temperature of a few hundred°K. These interstellar neutral hydrogen clouds (or *HI regions)* can be many light years across. A bright nebula has a similar density, but a temperature as high as 10^4°K. Most of the hydrogen gas in a bright nebula (also called an *HII region)* is ionized because of the presence of hot, blue stars. These young, short-lived stars are copious emitters of ultraviolet radiation with wavelength less than 0.0916 micron. Such short ultraviolet wavelength photons are capable of ionizing hydrogen in its ground-state.

Many articles on interstellar ramjets (Chapters 7,8) consider spacecraft accelerating in an "average" interstellar medium with 10^6 atoms/m^3. This is the density that would result if the high density "clumps" or nebulae could be averaged with the more diffuse material between them. The Sun resides in an "Intercloud Medium" or ICM. Knowledge of this component of the interstellar gas had to wait until astronomers were able to loft powerful ultraviolet observatories above the absorbing layers of the Earth's atmosphere. Two such platforms were launched as part of the NASA Orbiting Astronomical Observatory (OAO) program. Called *Galileo* and *Copernicus*, these craft provided the first long-term ultraviolet observing platforms above the Earth's atmosphere. Early OAO results and related ultraviolet observations have been summarized by Houziaux and Butler (4).

Lyman Spitzer, Jr. and Edward B. Jenkins have summarized ultraviolet observational results pertaining to the ICM (5). A typical ICM region has a neutral hydrogen density of 10^5 atoms/m^3, a proton density of 2×10^4 ions/m^3, and a temperature of $10^4\,^\circ$K. Partial ionization of the ICM may be due to interaction of interstellar atoms with low-energy cosmic rays and X-rays.

Technical
Note
11-1

Attenuation of Starlight

The extinction of starlight is caused by molecular scattering, large particle scattering, and absorption. We can express the fractional transmission, T, of a beam of starlight through distance, x, of the interstellar medium as:

$$T = \frac{I_x}{I_o} = e^{-\sigma x}$$

where I_o and I_x are the original and subsequent intensity of light passing through the interstellar medium, and σ is the attenuation coefficient, a function of wavelength λ, and the relative dimensions and material absorption characteristics of interstellar grains.

The attenuation coefficient can be expressed as a combination of terms:

$$\sigma = \sigma_m + \sigma_s + \sigma_a = \sigma_m + \sigma_e$$

where subscript m refers to molecular (Rayleigh) scattering, s to particle scattering, and a to absorption by particles. It is convenient to combine σ_s and σ_a into the *extinction coefficient* for particles, σ_e, a measure of the amount of incident light on the particles that is *extinguished*, that is, does not reach the forward direction.

Rayleigh scattering theory applies when the light wavelength is much larger than the size of the scattering particle—always the case for molecules. For molecules, Rayleigh scattering is essentially isotropic around the scattering center. The Rayleigh scattering coefficient for molecular scattering is:

$$\sigma_m = \frac{32\pi^3(n-1)^2}{3N\lambda^4} \qquad \text{for molecules}$$

where n = refractive index of the gas and N is the number of molecules per unit volume.

According to Hinds, the extinction coefficient for particles must be evaluated approximately in terms of the scattering efficiencies, Q_e, of particles in different size classes (i) (1):

$$\sigma_e = \Sigma \frac{\pi N_i d_i^2 Q_{e_i}}{4}$$

where N_i is the number concentration of particles with characteristic dimension, d_i. Q_e is in general a complicated function of particle size and shape, incident light wavelength, and the refractive index of the particle material. The size parameter, α, is very important in evaluating Q_e:

$$\alpha = \frac{\pi d}{\lambda}$$

According to Hinds, for $\alpha < 0.3$ (the region of Rayleigh scattering), Q_e is:

$$Q_e = \frac{8\alpha^4}{3} \left(\frac{m^2 - 1}{m^2 + 2} \right)^2$$

where m is the refractive index of the particle material.

Particles with larger size parameters are in a region governed by a theory originally developed in 1908 by Gustav Mie. Expressions for extinction efficiency as a function of α are enormously more complicated and have strong dependence on intricate complex variable functions of the angular direction from the scatterers. A complete discussion of *Mie Theory* may be found in the excellent references cited in (1).

Any interstellar expedition will have to be preceded by further probing of the ICM. The *in situ* measurements expected from Pioneers 10 and 11 and Voyagers 1 and 2 beyond the heliopause will be welcome additions to our meager store of knowledge regarding this realm. Of particular significance from spacecraft sampling will be information on small scale temporal and spatial variations in the ICM.

Although most of the matter in the interstellar medium is hydrogen, a fraction consists of more massive elements. A pioneer in the study of the more massive components of the interstellar medium was the Dutch astronomer Hendrick C. van de Hulst, who helped demonstrate that some interstellar matter must be concentrated as dust grains (6). Van de Hulst's doctoral research and postdoctoral work during the late 1940s found solid interstellar dust grains to be composed mainly of ice particles with an approximate size of 0.4 micron and a space mass density of about 10^{-21} kg/m^3. In an average part of the interstellar medium, the mean separation between adjacent dust grains will therefore be a few hundred meters. Van de Hulst had based his theoretical calculations on *Mie scattering theory* for light in the visible part of the spectrum, outlined in Technical Note 11–1. As more capable telescopes began to probe farther into the ultraviolet and infrared spectral ranges, carbon and silicates began to be recognized as constituents of interstellar grains.

By 1984, spectral absorption studies of the interstellar grains in the infrared between 2 and 10 microns had revealed that this material consists largely of complex organic (carbon-based) molecules. Most grains are found in the cold (10°K) environment of HI nebulae. A provocative and controversial result of the infrared absorption studies is the similarity between the absorption spectra of the grains and that of *E. Coli*, a terrestrial bacterium of similar size to the interstellar grains!

Radio telescope spectral-emission studies have, in fact, confirmed the existence of prodigious organic chemistry "factories" in cold interstellar clouds. More that 60 molecular species have been discovered so far in the interstellar medium, including carbon monoxide, alcohol, and formaldehyde (8). Most astronomers believe that the interstellar grains are prebiotic, not of biological origin. Perhaps life on the primeval Earth and similar planetary environments arose in an organic soup enriched by impacts of comets carrying interstellar organic grains, but most believe that life did not and could not have evolved in the cold, diffuse nebulae.

A small but vocal minority, however, claim that life not only evolved in the interstellar clouds, but also that these organisms—viruses, bacteria, even insects!—have actually played a direct and powerful role in terrestrial evolution. British astrophysicist Fred Hoyle has challenged astrophysical orthodoxy by publicizing the theory not only that life *continues* to originate in the dark interstellar clouds, but also that it *did not* originate on Earth (9)!

Hoyle and his very small group of followers believe that primitive life forms from the interstellar medium were deposited on the young Earth by impacting comets. Even today, they contend, many viral and bacterial plagues are due not to the migration and evolution of terrestrial organisms, but to alien organisms reaching the Earth's biosphere after a brush with a comet's tail. A similar view has been expressed by physicist and science-fiction writer Gregory Benford and David Brin in *Heart of the Comet* (10). This perspective is, of course, quite controversial. Most astronomers agree with the conclusion of R. E. Davies, A. M. Delluva, and R. H. Koch, that observational evidence does not support the existence of viral or bacterial life forms among the dark nebulae or comets (11).

Unlike many exotic interstellar speculations, however, Hoyle's theory might well be given a definitive test in the near future. Though resolution of this matter is not their objective, space scientists have their eyes on missions to collect a cometary sample and return it to Earth for analysis. Comets are considered to be remnants of the interstellar nebula from which the Sun and planets formed. Robot probes of several nations have already visited Halley's Comet and Comet Giacobbini-Zinner (1985–1987).

The study of the interstellar medium remains an active field. Gerritt Verschuur recently has described some tantalizing radio telescope observations of a quasar deep in extragalactic space that indicate the presence of small interstellar nebulae (12). These gas clouds, with sizes of about 7 AU or so and electron densities of $4000/cm^3$, might be more common than stars. If they exist, we may someday treat them as navigational hazards--or as starship fueling stations!

A discussion of the interstellar medium would be incomplete without mentioning those ubiquitous and ghostly messengers—the cosmic rays. Victor Hess's observations from seven free-balloon flights in 1912 first demonstrated to the scientific community that some gamma rays came from beyond Earth's atmosphere (13). In 1926, R. A. Millikan and G. H. Cameron coined the term "cosmic radiation" for this phenomenon (14). Later, physicists learned that positively charged atomic nuclei as well as electrically neutral gamma rays were in cosmic radiation. Walter Baade and Fritz Zwicky argued for a supernova origin of the nuclei in the cosmic radiation flux (15). Enrico Fermi demonstrated the method of cosmic nuclei acceleration by the interstellar magnetic field (16). Others, including Karl Otto Kieppenheuer and Vitali Lazarevich Ginzburg, discussed the possible connection between energetic charged cosmic radiation and galactic radio emission (17). Hannes Alfvén still maintains his belief that cosmic rays are local phenomena within the Milky Way galaxy (31).

The polarization of starlight provides evidence of the interaction between interstellar matter and magnetic fields. Interstellar dust grains act like small spinning magnets and thus align their short axes in the direction of the interstellar magnetic field, along the spiral arms of the galaxy. In the galactic neighborhood of the Sun, the field strength is 10^{-5} to 10^{-6} gauss.

The Starship and the Interstellar Medium

As discussed in Chapter 8, interacting with positively charged ions in the interstellar medium is an excellent way to decelerate a fast-moving starship. But what about the harmful effects: erosion of starships by dust particles and cosmic radiation bombardment of starship crews and scientific instruments?

Starship Erosion

When considering erosion by interstellar dust, it matters little whether the starship strikes a 10^{-16} kg silicon grain or a bacterium. The manner in which the kinetic energy of the dust grain is converted into heat in the starship's forward structure is what counts. Anthony Martin considered the interaction of interstellar material with a starship during his 1972 research on drag-screen deceleration (18). He considered an electrostatic drag screen, an electrically charged mesh that could function in a manner similar to the devices described in Chapter 8. At high relativistic velocities, drag-screen erosion would be due to sputtering or ionization caused by impacting dust grains, as was first suggested by E. T. Benedict in 1961 (19). Drag deceleration by the impacting dust grains would be negligible.

In 1972 N. H. Langton expanded upon the work of Benedict to calculate drag-screen erosion at relativistic velocities (20). He assumed conservatively that all the kinetic energy of oncoming dust grains would be converted into heat in the starship structure. At extreme relativistic velocities, an aluminum drag screen would rapidly erode by collisions with neutral matter. He proposed a wire-mesh deceleration screen as a superior alternative to a solid screen, to prevent erosion. And he suggested an ablation shield in front of the payload as a practical grain-impact absorber, even at relativistic speeds.

Conley Powell stated in 1975 that neutral interstellar hydrogen could be ionized by a very thin shield riding *ahead* of the ship (21). The material ionized on impact could then be trapped within a magnetic field. According to Powell, dust grains behave at relativistic speeds like blobs of free electrons and nuclei. Kinetic energy of the dust grain will be deposited along a track having the same width as the dust grain during its passage through a starship structure. He assumed that dust-vaporized material from the starship surface would behave like a perfect gas—an admittedly dubious assumption—and that vaporized material would be cooled by conduction with the surrounding solid material. At very high relativistic speeds (0.999 c), the time required for the gas to cool is a tiny fraction of the time required for the gas to escape. Erosion is therefore negligible. At lower speeds, Powell's theory indicates that somewhat more erosion may occur.

During 1977, N. H. Langton and W. R. Oliver reviewed previous work and concluded that graphite would make a better erosional shield than aluminum at relativistic speeds (22). In the 1978 Project Daedalus study, Martin reviewed and compared the erosional formulations of Benedict, Langton, and Powell (23). At the Daedalus probe's speed of about 0.15 c, Powell's method results in considerably less erosion than do the other methods. Considering more recent satellite observations of the ICM density, erosion may be less significant. In the worst case, a one centimeter thick shield is required for the half-century flight of the Daedalus probe to Barnard's Star.

In 1981 V. M. Bolie of the University of New Mexico examined relativistic flight in the even more diffuse *intergalactic* medium (24). The dominant contributor to drag in that medium is, surprisingly, the cosmic background radiation field—a remnant of the birth of the cosmos.

In 1986 there was an intriguing correspondence between Ian Crawford of the University of London Observatory and Robert L. Forward on the question of the erosion of laser light sails by interstellar dust grains (25). Forward believes that these sails are so thin that dust grains will pass through them without depositing much of their kinetic energy as heat. During a 10-ly journey at 0.2 c, only 1/500 of the area of a 0.0160 micron (160 Å or angstrom) thick light sail will be lost. However, Forward and we agree that a great deal of theoretical and experimental work on interstellar erosion must still occur before we can set off for the stars free of bad dreams.

Another interstellar navigational hazard was postulated by radio astronomer John Wolfe of the NASA Ames Research Center (26). As well as the more common 0.1 micron and lesser dust grains, it is possible that rare hailstone-size particles ($\approx$ 100 gram) exist in regions of interstellar space. Because the collision of such objects and a speeding starship would be catastrophic, and because *passive* protection via a massive forward shield would be prohibitive, *active* measures might have to be provided. Perhaps a forward-pointing millimeter-wave radar could be used to watch for these interstellar "golfballs." If one was found to be approaching, a high-power beamed energy device—a light or X-ray laser or a neutral particle beam—could be used to disintegrate or deflect the potential interstellar mine.

Cosmic Ray Protection

Detailed design studies of orbiting space habitats to be located near Earth have revealed that solar flares and cosmic rays would be serious threats to space colonists. A starship, of course, will be unaffected by solar flares during most of its flight, but it will require protection against cosmic rays. Most significant in the cosmic ray flux are heavier ions such as iron. A single relativistic iron nucleus could ionize atoms in millions of cells if it were to penetrate living tissue.

Nonreplaceable and nonreproducing cells such as those neurons in the spinal column are destroyed by ionizing radiation in space. During the 1969 Apollo 12 lunar mission, for example, the astronauts lost between 10^{-7} and 10^{-4} of their nonreplaceable cells because of cosmic rays. There is also the risk of cancers from this ionizing radiation, and to reduce those risks space voyagers must adhere to radiation-exposure guidelines. According to NASA special report (SP-413), adult radiation workers over the age of 18 are allowed a maximum of 5 rem/year (27). (One rem = 6.25×10^7 MeV [per gram of irradiated tissue] times the "quality factor" of the particular type and energy of ionizing radiation.) A member of the general population—especially children and developing fetuses—should not receive a dose greater than 0.5 rem/year.

Both passive shielding (layers of rock) and magnetic or electric field deflectors could be used to protect a starship population from galactic cosmic rays. Author Matloff reviewed the NASA study indicating that a passive areal mass density of 200–500 gm/cm^2 is sufficient to limit the cosmic ray dose on the ship's crew to between 0.5–5 rem/year (28). An interstellar population of 100 would require a shielding mass of 1.3×10^4 to 3.4×10^4 tons of rock. As pointed out by Paul Birch in 1982, electrostatic and electromagnetic shields will be able to reduce passive shielding mass requirements by several orders of magnitude (29).

The debate continues about what is a reasonable radiation threshold both for radiation workers and the general population. R. Silberberg and his colleagues have made a case for a space-operations limit of 35 rem/year (30). This level is achievable by an aluminum shield with a thickness of 9 cm, at vehicle speeds less than 0.05 c. At higher velocities, "active" electric or magnetic shields or supplementary passive shielding would be required to reduce the induced radiation dosage from interstellar protons and hydrogen atoms striking the ship's forward hull.

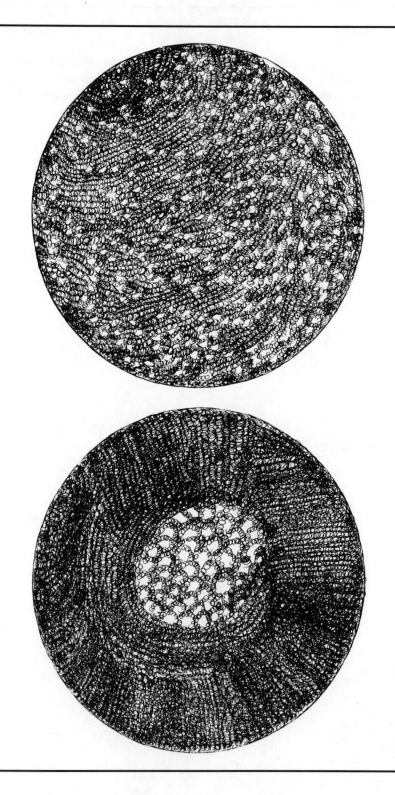

12 Starship Navigation and Visual Effects in Relativistic Flight

"Look," whispered Chuck, and George lifted his eyes to heaven. (There is always a last time for everything.) Overhead, without any fuss, the stars were going out.

Arthur C. Clarke, *The Nine Billion Names of God*

Navigating in interstellar space will inevitably recall the days of seafaring explorers who boldly set out for unknown lands, charting their courses by the Sun and stars. The objective of starflight, however, will be to *reach* those very stars, and make the Sun a mere point of light in the sky.

The Problem of Interstellar Navigation

Interstellar navigation is not easy, but it is orders of magnitude more tractable than starflight propulsion. It is fortunate that interstellar navigation, in contrast to propulsion, has a technology that is essentially in-hand. The necessary improvements in the current art of aerospace navigation are easy to foresee and to extrapolate. What remains is to select from many possible instruments and the available mathematical algorithms for position and velocity finding. There is one essential difference in high-velocity interstellar flight, however: significant relativistic optical effects arise that will alter the appearance and apparent positions of stars. Celestial navigators should not worry because, as we shall see, this is also a possible benefit.

Navigation, incidentally, is the art of determining one's position and velocity in some reference frame (such as an imaginary grid on the surface of the Earth), whereas *guidance* is the means of using naviga-

173

tional information to correct a trajectory to bring about some desired end—usually arrival at some predetermined destination with the proper velocity (often, but not always, zero). Any interstellar craft bent on rendezvous with a distant solar system will have to rely on auxiliary propulsion or fine vernier control of its main propulsion system to effect trajectory corrections that navigation data reveal are necessary.

Many kinds of navigation systems have been developed for terrestrial and interplanetary application. Some are not self-contained (autonomous) and rely on external human-engineered radio or microwave inputs, for example, geographically dispersed LORAN beacons for airborne, land, and sea navigation; and the GPS (Global Position Satellite) orbital position/velocity system for near-Earth navigation. Others, called *inertial navigation systems*, are autonomous up to a point. They use sensors—*accelerometers* and *gyroscopes*—and accurate clocks to establish a reference frame, with position and velocity information being maintained in a vehicle's computer by sophisticated mathematical techniques.

Since errors propagate with time in an inertial navigation system, it must acquire periodic *updates* from external navigation references, such as stellar fixes and environmental velocity measurements of various kinds. The modern approach to update the estimated position and velocity state of an aerospace vehicle employs the technique of *Kalman filtering*. This is a recursive mathematical algorithm for incorporating periodic externally derived position, velocity, or attitude measurements into *statistically optimal* estimates of the navigation state of a vehicle (1).

Human-engineered navigation beacons may have a place in interstellar flight, but they may be of limited utility—at least in the early days of starflight. For the first missions, it would be too much effort to seed navigation beacons among the stars. But a starship could derive some of its navigation data, for example, from properly time-coded signals in a microwave or laser beam coming from the Solar System and continuously focused on the starship aim point. Such a beam-riding system might be particularly appropriate for beamed power starships. A weak time-coded signal could be part and parcel of the power beam. Inertial navigation systems will also be of limited use because of the long time intervals during which errors could propagate and increase. The primary utility of inertial navigators in starflight will be to monitor transient propulsion correction maneuvers and changes in spacecraft orientation. But the most reliable and useful kind of interstellar pathfinding will be celestial navigation that incorporates various kinds of measurements of the surrounding starfield.

Technical
Note
12–1

Stellar Parallax Measurement

The apparent position of a nearby star shifts with respect to more distant background stars as Earth moves in its orbit. This angular parallax is conventionally defined for an observation baseline of one astronomical unit (AU) (see figure below). The geometric parallaxes of nearby stars are usually determined by careful measurement of photographic plates taken months apart which show the position of the star of interest and background reference stars.

Astronomers have, in fact, established a unit of distance called the *parsec*, which is a contraction of the words "parallax" and "arc second." A parsec is the distance to an object with a one arc second parallax shift observed from the ends of a one astronomical unit baseline, as in the figure. A parsec is equivalent to 3.261633 ly.

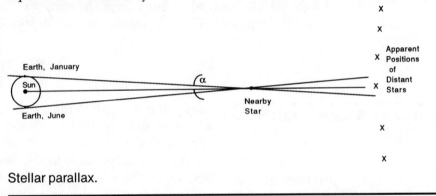

Stellar parallax.

Any celestial navigation that has ever been done within the Solar System has assumed the stars to be "fixed" on a sphere of effectively infinite radius. Most stars are so far away that the line-of-sight from an instrument within the Solar System to a star (sufficiently distant) will be parallel to the line-of-sight to that star from any other Solar System instrument. Moreover, during the short periods of typical terrestrial or interplanetary navigation, even the nearby stars have not moved, for all practical purposes. But interstellar navigation will be done, of course, in a three-dimensional manifold of moving stars—quite a different situation.

Astronomers have already done interstellar navigation in a limited way when measuring the tiny geometric *parallaxes* and proper motions of the nearer stars, observing their real and apparent motions against the background of more distant stars as time passes and as Earth cycles in its solar orbit. The geometric parallax of a star (for clarity in this example, a star *stationary* with respect to the Solar System) is its apparent angular shift on the background of much more distant stars caused by a shift in viewing location within the Solar System, for instance, Earth on opposite sides of its orbit (see Technical Note 12–1).

Determining the distances to the nearby stars by means of careful *astrometric* parallax measurements is equivalent to finding the Sun's position in a three-dimensional coordinate system. Measuring the *proper motion* of a nearby star—its motion perpendicular to the line-of-sight—is a matter of observing its position in the sky over a period of years and separating the oscillating parallax component of movement from the star's true linear motion. Knowledge of the star's distance from the Sun translates the star's angular motion to a relative linear velocity component, that is, in kilometers per second. A star's radial velocity—its line-of-sight velocity toward or away from the observer—can be determined from the Doppler shift in emission and absorption lines in its light spectrum.

Navigation Measurements in Starflight

Interstellar navigation must really begin long before a starship departs the Solar System. State-of-the-art astronomical techniques must be applied to define the positions and velocities of stars relative to the Sun with maximum precision and accuracy. Once the trajectory of the destination star is known, a tentative aim point in 3-D space can be established. During flight, the aim point may change slightly depending on the performance of the propulsion system. More refined data on the target's trajectory that may be developed from the starship, the Solar System, or the combination of the two perspectives, will also change the aim point.

The current levels of certainty in the positions and velocities of nearby stars are in flux as the techniques of astrometric observations improve on many fronts. Conventional parallax measurements of nearby stars are rather inaccurate, according to Wertz, who asserts that within 17 ly of the Sun the average probable error in distance to stars is 3%, and that there are only 700 stars with a distance measurement more accurate than 10% (2). These errors translate directly, of course, into similar percentage inaccuracies in stellar proper motions. But astronomer George Gatewood and his colleagues at the Allegheny Observatory have recently developed electronic astrometric techniques (see Chapter 16) that may improve those figures by a factor of 5 to 10—routine parallax measurements accurate to 0.001 arc second (3).

We have not seen the end of accuracy improvements in astrometry. So-called "long baseline optical interferometry" (again, see Chapter 16), in principle, has no limit to the accuracy of its stellar position measurements. Some specialists in this blossoming field anticipate 10^{-6} arc second accuracies for arrays of optical telescopes based in space (perhaps on the Moon) early to mid next century (4). This implies a distance resolution (transverse to the line of sight) of about 0.00003 AU or only 5000 km for stars 100 ly away! It is also true that sensitive radio telescopes linked together in VLBI (Very Long Baseline Interferometry) arrays have recently been able to observe some microwave-emitting nearby stars with 500 micro-arc second resolution and milli-arc second accuracy position measurements of some stars have also been obtained (5). For stars 30 ly distant, VLBI parallax measurements with such accuracy would translate to line-of-sight distance determinations of only a few hundred AU.

As a result of efforts to detect extrasolar planets (Chapter 16), spectroscopic techniques of measuring the radial component of a star's velocity have also made amazing strides in recent years. It is now possible to obtain measurements with better than 10 m/sec accuracy.

So before a starship ever takes off, we should know exactly in which direction to send it to rendezvous with a nearby star. Tracking a departing starship's directional microwave beacon with a long baseline array of radio telescopes should be adequate to assess the direction of its trajectory and the craft's distance from Earth with very high accuracy. For a slowly moving interstellar ark or robotic probe ($<< 0.1$ c) there would be more than enough time to radio trajectory correction commands from the Solar System. For higher-velocity starships, autonomous astrometric navigation systems would be required because limited time would be available to radio course correction commands from the Solar System.

Two engineers who worked on the Apollo navigation and guidance system, David Hoag and Walter Wrigley, published one of the most comprehensive discussions of interstellar navigation just as the Apollo program was winding down (6). Strongly influenced by the stellar update inertial navigation system used on Apollo, the authors outlined an interstellar navigation system that would optimally mix stellar observations with outputs from an inertial navigation system. They concluded: "The technology of navigation and guidance demonstrated in modern autonomous space systems, which have been adequately served by a Newtonian concept of the physical world, will not require any fundamental modifications in supporting an interstellar mission in which relativistic effects will necessarily play a prominent role."

The *navigation state vector* of a starship will, of course, include three components describing position and three elements specifying velocity, perhaps in a coordinate system centered in the Sun. The angles characterizing vehicle attitude will also appear in the state vector. And since a good deal of navigation information will derive from star measurements, a significant part of the navigation state vector will serve to characterize the positions and velocities of selected stars.

Apart from straightforward measurements of the apparent directions to navigation stars, other stellar observations could be useful in various navigation schemes, as suggested in references 2,6,7,8, and 9:

1. *Apparent direction to distant stars, external galaxies, and quasars*: The apparent direction from a starship to distant astronomical objects provides information not only on the orientation of the craft, but also on its velocity, through an effect called the *aberration* of starlight. To make best use of these measured directions, however, the reference objects should be so distant that they have negligible geometric parallaxes. Stellar aberration, which provides a handle on the starship's velocity, is the tiny apparent shift in the direction of a star because of the finite velocity of light, an effect akin to the shift in the direction of rainfall apparent from a moving vehicle. The effect is small at low speeds, but at high speeds it grows much larger and markedly distorts the appearance of the starfield (see section below and Technical Note 12–2).

2. *Apparent direction to nearby stars*: Measuring the direction to nearby stars from a starship and combining that information with an onboard three-dimensional mathematical model of their positions (taking into account the stars' changing relative locations) is equivalent to measuring the geometric parallaxes and hence distances to those stars. This provides data to fix the starship's position. The "baseline" for the changing apparent star directions can be drawn out by the starship itself as it progresses on its flight path. Alternately, a long baseline could be created by auxiliary observing stations periodically dispersed several AU from the starship, which would report apparent stellar directions simultaneously.

3. *Angles between stars*: A variant of measure (2) that would possibly be easier to implement because it eliminates the need for a highly accurate stable platform on the starship from which to reference star directions. A desirable geometry: one star nearby and the other far away and without significant parallax.

4. *Star brightness*: It is possible to select stars that are likely to have very stable inherent brightness for the duration of a mission. Determin-

Technical
Note
12-2

Stellar Aberration

Stellar aberration, an effect much bigger than geometric parallax observed from the Solar System, has been known since the early eighteenth century. British astronomer James Bradley found that the velocity of Earth in its orbit was causing stars to describe tiny elliptical paths in the sky (see figure below). The major axis of such an ellipse subtends an angle (in radians) of $2V/c$, where V is the velocity of Earth in its orbit and c is the speed of light. The minor axis of the ellipse is $2(V/c)\sin\theta$ where θ is the angle of elevation of the star above Earth's orbital plane. Since Earth's orbital velocity is about 30 km/sec, stellar aberration seen from Earth amounts to a maximum annual shift of 2×10^{-4} radian or about 41 arc seconds.

At higher velocities, the transformations of relativity must be taken into account and lead to a formula for *relativistic stellar aberration*:

$$\cos\theta_1 = \frac{\cos\theta + \dfrac{V}{c}}{1 + \dfrac{V}{c}\cos\theta}$$

where θ_1 is the apparent angle of a star from the velocity vector of the starship and θ is the apparent angle to the same star when the starship is at rest relative to the star. Knowing θ and measuring θ_1 is therefore a way of measuring starship velocity relative to a reference star. This formula of course reduces to Bradley's non-relativistic formula for sufficiently small V.

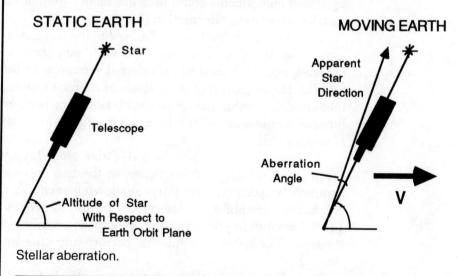

Stellar aberration.

ing photometrically the apparent change in light intensity from such stars, would—by the inverse-square law—be a corroborating measure of their distance. But note well, particularly in high-speed starflight, that the intensity of starlight measured is for light that left the star years earlier. Therefore, this and all other methods of determining distances to nearby stars does not provide an "instantaneous" measure of star-to-starship distance. Star spots, flares, and other unpredictable intensity-altering phenomena could also reduce the utility of this measurement.

5. *Spectroscopic Doppler shift*: The Doppler shift in the characteristic emission and absorption lines of a star will provide a measure of the starship's velocity component in the direction of an observed star. Looking out in the direction of flight, the shift will of course be toward shorter (bluer) wavelengths, while looking at receding stars will find them with a red-shifted appearance. Doppler shift techniques should ultimately be able to provide relative velocity measured to an accuracy of a few meters per second or less. Since spectral frequency shifts would also give rise to changes in apparent stellar brightness, Doppler shift measurements would have to be used to correct distance determinations obtained from brightness measurements (see Technical Note 12–3).

6. *Angular size of stars*: Observing the diameter of stellar disks by optical interferometry could also serve as a stable distance measurement to reveal starship position. Alternately, a starship navigator equipped with an accurate orbital description of a binary star system with visibly separated components could measure their apparent angular separation and derive position information (7).

7. *Times of eclipses of eclipsing-binary stars*: Markowitz has suggested that the times of eclipse of close binary stars, so-called eclipsing binaries, could be used as an external reference in interstellar navigation (9). He suggests that if the times of eclipse of three such binaries are measured, a starship navigator could determine her position, and if four binaries were observed in eclipse she would get an independent measure of time as well.

Eclipsing binaries have a period of about one day and display regular sharp rises and drops in brightness as the two components circle their common barycenter, alternately shadowing each other. Markowitz calculated that the ability to determine the minimum of an eclipse to within 1 to 0.1 seconds implies a position measuring accuracy of several hundred thousand kilometers—wonderful performance for interstellar navigation.

8. *Pulsar signals*: The intense magnetic fields of rapidly rotating neutron stars give rise to regular bursts of radio energy that make such

Technical Note 12–3

The Relativistic Doppler Effect

The observed shift in the frequency, ν, of electromagnetic radiation emitted in one frame of reference as measured by another frame of reference moving at velocity, V, *away* from it is by the relativistic Doppler effect:

$$\nu_{obs} = \nu \sqrt{\frac{1 - \frac{V}{c}}{1 + \frac{V}{c}}}$$

If the emitting and observing frame are *closing* on one another rather than separating:

$$\nu_{obs} = \nu \sqrt{\frac{1 + \frac{V}{c}}{1 - \frac{V}{c}}}$$

There is also a so-called *transverse* Doppler effect for radiation that approaches an object at angle, θ, to the direction of its motion:

$$\nu_{obs} = \nu \frac{\sqrt{1 - \frac{V^2}{c^2}}}{1 - \frac{V}{c}\cos\theta}$$

Note that this formula reduces to the two above expressions for the appropriate angles, $\theta = 180°$ and $\theta = 0°$ respectively.

stars worthy of the name pulsars. Sometimes the pulses are only milliseconds apart, and the pulse time signature of a neutron star serves as it unique identifier. With an accurate time base aboard the starship, pulsar signals could be employed in the same way as eclipsing binary stars to determine starship position. The Doppler shift in the apparent pulse rate would also provide a measure of relative starship velocity.

Other external navigation information may come from a pre-arranged coded navigation beacon beamed from the Solar System. Another source of navigation information, velocity measurements in particular, may be observations of the microwave cosmic background radiation, which defines what is, in effect, the frame of cosmic rest. Such microwave measurements have already been used to determine the velocity vector of the Sun in interstellar space.

Starscapes at Relativistic Speeds

At small fractions of the speed of light, the view of star-studded space from an interstellar craft would seem unremarkable except for the gnawing absence of a bright nearby sun and its planets. But as a capable starship accelerates to relativistic and then extreme relativistic speeds, the view of the surrounding starfield changes in an unusual way. Many who have calculated from Special Relativity what the "view from the starship bridge" would be like have projected: dramatic compression of the entire starfield toward the direction of travel caused by stellar aberration; startling changes in star colors due to the Doppler shift; and the extraordinary brightening of stars, also due to the Doppler shift (10–18). As noted in Chapter 3, relativistic phenomena are typically on the order of "15% effects" even at 0.5 c, so dramatic visual effects do not occur until still higher velocity is reached.

Contemplating the view from relativistic starships began in earnest in the early 1960s with the works of Rytov, Sänger, and Oliver. Oliver's aptly titled paper, "The View from the Starship Bridge and Other Observations," is one of the classics. It is a tribute to Bernard Oliver, a prominent researcher in the SETI field (who does not believe in the practicality or desirability of starflight) that he has made fine analytical contributions to the study of hypothetical interstellar flight. Moskowitz and Devereux, in particular, were early pioneers in using digital computers to portray the effects of stellar aberration on real starfields for imaginary flights at relativistic velocity in certain directions away from the Sun (13–15). Their simulated celestial scenes clearly show the forward direction compression of the starfield.

The culminating works of starscape simulation and analysis, however, are the two amazingly thorough studies done by Stimets, Sheldon, and Giles in the early 1980s (16,17). For the first time, such simulations accounted for the significant content of normally invisible ultraviolet (UV) and infrared (IR) radiation in stellar spectra. At high velocities, IR radiation forward of the starship and UV radiation rearward can be Doppler-shifted into the visible spectrum and radically change the visual pattern of stars—apart from the already severe effect of stellar aberration and, over time, geometric parallax. These spectral shifts create uneven brightness variations throughout the starfield, seeming to bring new stars into view and make others dim and disappear. The authors showed that forward direction image compression is significant above 0.5 c and that over 0.6 c "the forward view is comprised mainly of blue-shifted red giant stars."

Moreover, as others found earlier, the forward "cone of clustering" develops a more and more acute angle as speed increases. Also, an enormous "cone of blackness" forms rearward except at the singular, directly astern point, $\theta = 180°$. At sufficiently high velocity, the once uniform (isotropic) cosmic microwave background radiation becomes enormously concentrated and bright. Stimets and Sheldon vividly describe the wildly distorted picture of the universe in extreme relativistic flights:

"Ultimately, at the very highest attainable speeds, the all-pervading 2.7°K cosmic background radiation would be blue-shifted into the visible region and reach its brightest intensity in a forward cone spanning less than one minute of arc [1/30 apparent Moon diameter], with a visual magnitude equivalent to about one-tenth the brightness of the Sun as seen from the Earth, with the remainder of the sky completely black. The view of the Universe in this extreme would thus dwindle to a mere pinpoint of light, having dazzling brilliance with its surround of utter darkness. Only the spacecraft itself and its occupants would retain the familiar span of space and time. Upon deceleration, environmental space and time would revert to being in synchrony with that experienced by the astronauts. The initially all-embracing Universe would have seemed in the course of a single flight to have become crowded forward into a brilliant minute lodepoint of light and then, like an unfolding rosebud, to have opened up again into its full majestic breadth, and 'natural' color."

As calculated by Stimets and Sheldon, views of the forward direction (60° cone) starfield in a flight toward the north celestial pole are portrayed in Figure 12.1 for starship velocities from 0 to 0.992 c (16). The authors' computer program, CELESTE, can generate such scenes automatically and display them on color graphic video displays. Romantic science-fiction writers please note: Stimets and Sheldon's precise calculations refute the claims made by some earlier investigators that a "starbow" would appear in the forward direction—multicolored rings of stars, with a darkened central spot. Stimets and Sheldon note that the population of apparent bright red stars increases more rapidly than bright blue stars. At velocities near 0.9 c the visual magnitude of stars in the forward direction would be comparable to the planet Venus at times of its maximum brightness in terrestrial skies. In contrast to the approx-

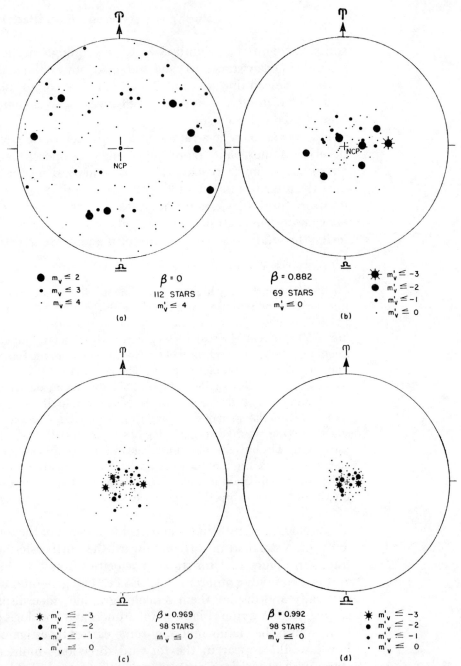

Figure 12.1 Starscapes at Relativistic Velocities. Starship motion is in the direction of the north celestial pole (NCP). The parameter, β, represents the starship's velocity as a fraction of the speed of light, that is, $\beta = 0$ for a stationary vehicle. The parameter, m_v, represents visual magnitude—increasing brightness with decreasing m_v, as is conventional. (Courtesy R. W. Stimmets/E. Sheldon and JBIS)

imately 5000 stars ordinarily visible to the unaided eye from Earth, 100,000 would then be in view scrunched up in the forward direction.

It is absolutely clear that the extraordinary appearance of starscapes at high relativistic speeds would have profound consequences for interstellar navigation, to say nothing of its psychological impact on star voyagers. The severe effects of stellar aberration and Doppler shift at extreme velocities would transform the linear problem of low speed celestial navigation to a highly nonlinear one critically dependent on high-precision astrometric measurements.

The potential effects that the contorted view of space would have on the psyches of starship inhabitants is worrisome, perhaps the first *direct* human encounter with the alien realm of spacetime distortions. Would the voyagers develop a kind of "cosmic claustrophobia" ("C-sickness"—a marvelous term offered by Stimets and Sheldon) or disorientation as they witnessed their once familiar universe squashed beyond recognition? If reactions to past supposed barriers are any guide, for instance, "high-speed" travel by rail in the era of horseback transportation, supersonic flight, and weightlessness, human beings will adapt with ease to the bizarre vista from the relativistic starship bridge.

13 Starflight Between Fact and Fancy

It is difficult to say what is impossible, for the dream of yesterday is the hope of today and the reality of tomorrow.

Robert H. Goddard

Any sufficiently advanced technology is indistinguishable from magic

Arthur C. Clarke's "Third Law"

Wormholes in Spacetime

A mad inventor, ignored and abused by the world, discovers in gloomy isolation fantastic new insights into the laws of nature. In a popular science-fiction theme, the tormented genius works alone and from his (rarely *her!*) well-equipped laboratory, a wondrous machine emerges. Unlike the starships described elsewhere in this work, the craft is no $100-trillion speedster able to reach Alpha Centauri in 50 years or even a $500-billion "slow boat" that requires a millennium for the crossing. The inventor's vessel is comparatively inexpensive. Although he has put his life savings into the starship, the inventor is no billionaire. Constructed using surplus Air Force or NASA equipment, the ship is no larger than a Winnebago.

A small gathering of well-wishers huddle expectantly as a friend of the genius anoints the ship's bow—or front fender—with a bottle of champagne. His brief speech over, the inventor smiles for the lone press photographer, enters the ship, and flips a switch. With a flash of light and a clap of thunder, the bargain-basement starship vanishes into "hyperspace." The craft reemerges in normal space near Alpha Centauri. Encountering many hardships—malfunctioning equipment, nasty aliens, uncharted comets, and asteroids—the explorer perseveres. Ultimately, having opened the cosmos to humanity, he returns to Earth and a hero's welcome.

Fiction—yes, romantic drivel—perhaps: But similar Promethean fantasies have inspired generations of adolescents. Some of them have grown up (or have they?) to become serious researchers. What physics—

187

if any—lies behind these tales? Is there perhaps some loophole in physical theory that we have not stumbled across, much as Newtonian mechanicians overlooked relativity for centuries? Might an entirely new physics emerge that will enable us to leap across the light years with the same abandon that we now cross the oceans?

Fact: The vast majority of space scientists and engineers are far from optimistic about the prospect of finding a hidden, easy pathway to the stars. Without hard evidence, they are reluctant to consider major modifications to the known laws of physics. Only the most open-minded among them will seriously speculate along these lines.

One of the most daring speculators in the ranks of serious scientists has been physicist Robert L. Forward. In 1974, Forward delivered a lecture to the annual meeting of the Science Fiction Writers of America in Los Angeles. Forward's "Far Out Physics" describes a number of hunting grounds for researchers seeking antigravity or faster-than-light space drives (1). A recent expansion of these ideas is his book, *Future Magic* (2).

One "far out" space drive would rely on *substellar black holes*—mini–black holes formed not in the "conventional" way by the collapse of massive stars but by hypothetical processes at the birth of the universe. Allowed by Einstein's General Theory of Relativity and the supporting work of Stephen Hawking and others, a black hole—a radical distortion in the fabric of space time—occurs when the density of matter becomes too great and a self-accelerating process warps four-dimensional space-time into a *point singularity*. The singularity is surrounded by an *event horizon*, a spherical boundary that imprisons all mass *and* radiation that happens to cross it—hence the adjective black, because the hole does not emit radiation. A black hole nonetheless has mass, angular momentum, and charge.

Small black holes with masses from micrograms to the mass of asteroids or large mountains may lurk in the barren reaches of space, in the cores of stars, or even at the centers of the planets. A mini–black hole with the mass of an asteroid has an event horizon the diameter of an atom. As Forward suggests, "There could be swarms of them in the centers of the Sun and planets, slowly eating them up an atom at a time."

If we could only detect and trap some of these tiny, massive critters, we could well use them in all kinds of space-drive research. Forward has proposed a "mundane" use: as "catalyst" for fusion reactions, a mini–black hole being able to create enormous plasma densities in its vicinity. Science-fiction writer James Hogan was clearly sufficiently impressed with mini–black holes to include them in *The Gentle Giants of Ganymede* (3). His Minervian starship *Shapieron* is propelled to extreme relativistic velocities by a black-hole engine. Under certain conditions, a large fraction of the mass falling into a black hole could, in theory, be converted into propulsive energy.

We should emphasize that observational evidence for the existence of mini–black holes is nil, although most astrophysicists accept the substantial evidence for black holes with stellar masses or larger: relics of supernovae explosions, the cores of galaxies, and so on. And even if one could locate mini–black holes *within* the Solar System, one would still have to contend with the difficult problem of how to domesticate them—store them stably within a spacecraft and contend with their "evaporation" as *Hawking radiation.*

Although too massive to be useful in starship engines, black holes of stellar mass may have applications to starflight. Black holes of stellar mass are thought to be produced by the collapse of stars greater than 1.4 to 2.5 times the mass of the Sun, occasionally in titanic supernova explosions (for stars of more than 4 to 6 solar masses). Soviet astrophysicist N.S. Kardashev and others have considered the spacetime warping of these strange denizens of the cosmos (4). Kardashev envisioned the galaxy linked by a black-hole "subway system," with ramjet propelled spaceships flitting between solar systems and the nearest "black-hole station." These spaceships approach a black hole along certain well-defined critical trajectories, leave normal spacetime, and emerge elsewhere or "elsewhen." In *The Cosmic Connection*, Carl Sagan dubbed black holes "cosmic Cheshire cats" (5). In his first science-fiction novel, *Contact*, Sagan drew an even closer analogy between such a black-hole spacetime network and present urban subway systems—minus the graffiti!

What about those unfortunates who chance to live near a star a hundred light years or more from the nearest black hole? Are the poor souls of the galactic boonies doomed to be ignored forever by the cosmopolitan civilization evolving around them? Not necessarily, says Adrian Berry in *The Iron Sun* (6). Using *in reverse* the ramscoops discussed in Chapter 8, he proposes that a civilization might "herd" interstellar gas over enormous volumes, compact it, and create an artificial rotating black-hole "subway stop."

Although the last word on the transgalactic hyperspace subway has yet to be written, Cornell University astrophysicist Thomas Gold has poured cold water over the hopes of its enthusiasts (7). According to his calculations, an astronaut or spacecraft approaching a stellar mass black hole would be stretched catastrophically by severe gravitational tides

The Tides of a Black Hole

Using Newton's law of gravitation compute the tidal force on a spacenaut's body near a black hole of one stellar mass:

$$F_{grav} = \frac{GM_oM_s}{R^2}$$

where G = the gravitational constant, 6.67×10^{-11} N m²/kg²
  M_o = object mass
  M_s = star mass
    R = separation between the center of the hole and the center of mass of an object near it

Differentiating with respect to R:

$$\frac{dF_{grav}}{dR} = \frac{-2GM_oM_s}{R^3}$$

As the object changes its position by ΔR, the change in gravitational force is:

$$\Delta F_{grav} = \frac{-2GM_oM_s\Delta R}{R^3}$$

For a spacenaut 1 km from the center of a solar-mass black hole, oriented with feet toward the hole as though about to stand on it, calculate the differential gravity force or "tide" between the spacenaut's head and feet, assuming 0.1 body mass in the head and 0.1 body mass concentrated in the feet. For the following values: M_o = 1/10 astronaut mass = 10 kg; solar mass = 2×10^{30} kg; ΔR = 2 meters; and R = 1000 m, the force between head and feet is about 6×10^{12} Newtons, or a tidal acceleration of 6×10^{10} g!

(Technical Note 13–1). The ship's occupants might find themselves, at least temporarily, with the dimensions of a spaghetti noodle—distinctly uncomfortable to a crew member, not to mention embarrassing for her spacesuit designer!

Dean Drives and Other "Impossibilities"

A few stops beyond black holes comes *antigravity* in the lexicon of "magic" spacedrives. Antigravity for space propulsion seems to have caught the attention of fewer people than have black holes. Like perpetual motion machines, antigravity devices are usually given short shrift by patent offices as well as by physicists and engineers. Yet some serious scientists have taken time to examine some of the extraordinary claims. Robert L. Forward, for one, never tires of at least *listening* to the latest "crackpot" proposal, hoping to gain some inspiration or new serendipitous idea from it. Forward has also developed quite a few interesting exotic possibilities for starflight propulsion on his own, in addition to his more "conventional" ones: the laser-pushed sail and Starwisp.

Forward has considered a group of antigravity machines, among them a so-called Special Relativistic and a General Relativistic Antigravity Machine (1). The former is a torus with a pipe wrapped around it in the manner of wire around a circular electromagnet core. But instead of conducting electricity, the pipe circulates an ultradense accelerating fluid that in theory might cancel ambient gravity fields. His General Relativistic antigravity machine would be like a very dense smoke ring turning itself inside out (as smoke rings usually do), which Forward claims would create the equivalent of an artificial gravity field, which again could be used to cancel ambient gravity fields. Even if advanced technology could bring these machines about, it might be argued, "So what, how is this going to propel us? Gravity is not the problem in interstellar flight." The argument is correct, up to a point, but think of the kinds of astronomical engineering—disassembly of planets and asteroids, and so forth—one could carry out with such machines, and thus help to do some of the projects mentioned in Chapter 9.

Forward is particularly fond of the concept of *negative matter*, not *anti*matter, mind you, but *negative* matter. (Forward's most recent and detailed account of this idea is in his paper, "Negative Matter Propulsion," AIAA 88–3168, presented at AIAA/ASME/SAE/ASEE 24th Joint Propulsion Conference, Boston, MA, 11–3 July, 1988.) Negative matter would *repel* instead of attract all other matter. A negative and positive mass placed together would automatically accelerate and not stop! As Forward says, "The negative mass would keep pushing the positive mass and the positive mass would keep attracting the negative mass and they'd keep on going." Forward and physicist Herman Bondi have shown that apparently the concept of negative matter does not violate some of the basic laws of physics. In particular, negative matter seems not to violate the conservation of energy or the conservation of momentum. But whether it does not violate some other principle is unknown (the always problematic Second Law of Thermodynamics, perhaps?). And how could we create negative matter in the first place? Forward says that in principle one could "take empty space and rip out the negative mass and positive mass at the same time."

If negative matter is not bizarre enough, how about "inertia control"—finding some loophole in physical theory that would allow us to reduce the inertial mass of a body, thereby increasing the acceleration achievable for a given applied force? What about violations of Newton's Third Law to make a "reactionless space drive." Engineer G. Harry Stine has actually participated in experimental investigations of alleged reactionless space drives. Writing in *Analog*, Stine described a research project conducted during the early 1960s to test one such device (8).

Stine's interest in anomalous space drives stems from the influence of the former editor of *Analog*, the late John W. Campbell, Jr. In the late 1950s and early 1960s, Campbell published a number of articles in *Analog* about the apocryphal *Dean drive*, a device with the supposed ability to convert rotary motion *inside* a closed container into net linear motion of the entire system.

If Norman Dean's device, which was awarded US Patent # 2886976, works as claimed, the laws of physics are in need of drastic revision. Constructed of counterrotating masses in a framework able to alter the masses' center of orbital motion in a controlled manner, the Dean drive appeared to Campbell and others to convert self-contained rotary kinetic energy into linear momentum. A "Dean drive solar system" could suddenly change its direction of motion by judiciously arranging the motion of the planets!

To say the least, a working engine of this type would have enormous implications, not the least of which would be more efficient high-speed spaceflight. Skyhooks might be possible that would permit massive buildings to be moved with ease from site to site. Architects might have a field day with "flying cities." The decaying civil engineering infrastructure of urban America would no longer be of concern because motor vehicles would no longer need wheels. They would simply float above the ground like the antigravity sleds in a Star Wars movie.

Working with William O. Davis, retired USAF Colonel and former director of the Air Force Office of Scientific Research, New York University Physics Professor Serge Korff, and others, Stine investigated the Dean drive and other anomalous devices. In an attempt to develop at least a theoretical framework for space drives, Davis proposed a modification to Newton's Second Law containing terms that might explain the alleged forces (Technical Note 13–2). It is noteworthy that in at least one commonly used college physics text, the validity of Newton's action-reaction Third Law for long-range interactions is brought into question, over the issue of the possible non-simultaneity of the action and reaction (9).

An intuitive way of comprehending how a Third Law-violating de-

William Davis's Proposed Modification of Newton's Second Law*

In Davis's Mechanics:

$$F = kx + V\frac{dx}{dt} + m\frac{d^2x}{dt^2} + Dm\frac{d^3x}{dt^3}$$

where x = distance, t = time, k = a constant akin to the Hooke's Law spring constant, V = constant akin to a viscous damping coefficient, m = mass, and D = "critical action time."

The third term in this formalism is, of course, Newton's familiar Second Law of motion. According to Davis, we are usually not aware of terms 1, 2, and 4 of his proposed force law, because under most conditions these are very much smaller than the third term. However, we again emphasize that experimental evidence for this and other proposed modifications to the fundamental laws of mechanics is so far nonexistent.

*Take with a very large grain of salt.

vice might work: When you push against a solid object with your hand, we normally assume that the object pushes back *instantaneously* against the hand with a force opposite and exactly equal to the imposed force. But imagine instead that the hand creates a pressure wave in the solid atomic lattice of the object. The wave penetrates a certain distance into the material before a "rebound" occurs. There might be a finite lag time—perhaps measured in micro- or nanoseconds—between the imposed force and the reaction force. Now, if the object were to rotate 180° during the time between the action and the reaction force, and if the rotation required zero-energy expenditure (a difficult trick, indeed!), action and reaction forces would be of the same sense and produce a net force in one direction!

All these approaches to a new physics, though stimulating and thought provoking, are ultimately meaningless if experimental validation is not forthcoming. Stine wrote that he "felt" the push of the Dean drive against his hand, but he lacked the means to quantitatively test the device. So he developed a simple free-hanging pendulum (Figure 13.1) to test alleged space drives for unidirectional forces. Although Norman Dean did not consent to a pendulum test for his invention, Stine and his associates examined a large number of other claimed space drives. *Not one* demonstrated a unidirectional force when suspended from the pendulum.

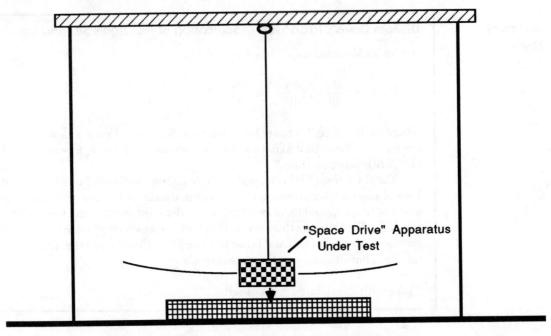

"Space Drive" Apparatus
Under Test

Figure 13.1 Pendulum test for suspicious stardrives.

The reluctance of an inventor to put a machine—in which she has invested perhaps hundreds of thousands of dollars—to the ultimate test is understandable. The failure of a critical component might doom her efforts to be believed. However, scientists and engineers have also invested great time and effort in winning confidence in that framework of knowledge we call physical law. If someone claims to have found a violation of that framework, the onus of proof is rightfully borne by the claimant, not the defenders of physics! That is the bottom line for the Dean drive and other improbable space drives.

Superluminal Travel

Putting aside the prosaic dynamics of unidirectional space drives, let us move on to the scholastic question: Is the speed of light barrier sacrosanct and inviolable? Is superluminal flight remotely possible? Arguments given by Paul Birch, H.D. Froning, Jr., and R.T. Jones suggest that neither

Special nor General Relativity absolutely rules out supralight velocities (10–12). Their many arguments are very appealing and creative, but in no way *prove* that faster than light travel is possible. They merely establish that relativity theory and causality may not be violated by certain potential loopholes, for example, higher dimensionality of spacetime or hypothetical *tachyon* particles that may *always* travel superluminally. Alas, not a shred of observational or experimental evidence exists for faster than light travel. Their claims are simply unsupported. Besides, with a black-hole subway system, who needs superluminal starflight?

More recently, physicists Michael S. Morris, Kip S. Thorne, and Ulvi Yurtsever of the California Institute of Technology investigated a more detailed mechanism by which advanced civilizations might maintain spacetime "wormholes" (20). They claim tentatively to have demonstrated the possibility *in principle* of superluminal travel between widely separated parts of spacetime, and that this would lead to causality violations and the prospect of time travel. The researchers suggest that a spacetime wormhole, the by-product of black hole formation, could be maintained by employing specially placed charged spheres. One way of creating such a wormhole, they say, is to pluck out and amplify to macroscopic size a microscopic fluctuation in the universe's presumed underlying quantum mechanical "spacetime foam."

One way that a starship crew could *effectively* break the apparent cosmic speed limit is to travel "conventionally" at moderate sublight speed while in suspended animation (see Chapter 14). Arriving at the destination, the crew would wake up and activate their handy F. J. Tipler time machine and enter a long-past era! All they need to pull off this feat of time stasis is a very dense cylinder spinning at about half the speed of light (18)!

The Quantum Ramjet

Froning has also dreamed up another intriguing approach to starflight, one that is more palatable than faster-than-light travel—at least it has some basis in physical theory (13,14). His "quantum starship" relies on the well-established theory of quantum fluctuations in the energy of the vacuum. These occur throughout the universe on extremely small scales of time and distance. Over times on the order of 10^{-15} second and lengths of about 10^{-33} centimeter, masses as high as 10^{-5} grams and energies as large as 10^{16} ergs pop in and out of existence interminably. Conventional physics supports this picture, but it is an entirely different matter to make use of vacuum quantum fluctuations to propel a starship. Froning's quantum ramjet would work by "ingesting" the energy of the quantum fluctuations and converting it to propulsive energy. If the quantum starship could tap only a very tiny fraction of the theoretically available mass/energy of the vacuum, it could accelerate rapidly to relativistic velocities.

The philosophical implications of the quantum ramjet are startling. If, as the *new inflationary theory* of cosmology mandates, the universe evolved from a quantum fluctuation that somehow grew to its present enormous scale, the same thing might occur as a matter of course in the quantum ramjet (15). Would the quantum ramjet create and destroy countless universes as it travels our cosmos, and would the ramjet crew be as gods to the countless beings in the universes that would support their flight?

Consider now the quantum gravitational shield proposed by J.M.J. Kooy (16). If it were possible to modify the *direction* of interaction of a spacecraft with gravitons—the hypothetical quanta responsible for gravitational force—one could use directionally variable gravity to push or pull a spacecraft. High relativistic velocities would be possible, and the universe itself would do all the work! The "gravity planar" used by the Kzin in Larry Niven's story "The Warriors" may be an example of a directional gravity drive. The cocky and overconfident Kzin are defeated by a human crew riding a far more primitive laser ramjet!

Cosmologists and observational astronomers have established in recent years convincing arguments that from 90 to 99% of the universe may consist of a yet to be identified, invisible or dark matter, unlike ordinary matter except that it interacts gravitationally with it. So what we normally think of as the main constituents of the universe—luminous stars, galaxies, planets, and interstellar gases—may be nothing but the frothy crests of a deep ocean of dark matter. To say the least, it is premature to consider using this invisible, pervasive substance in any kind of propulsion scheme. But after physicists have identified what seems to be the bulk of the universe, would it not be wonderful to begin thinking about "dark matter ramjets"?

Now if these speculations are too prosaic, how about a quantum-jump/psionic starcraft? Even work-a-day physicists know that electrons instantaneously jump from one energy state to another in an atom. What if we could master the control of such quantum jumps, apply them on a macroscopic scale and jump instantaneously between the Earth and the surface of planet Tau Ceti-4? Sounds suspiciously like telekinesis, the alleged extrasensory phenomenon—probably rubbish—in which the "force of will" moves objects at a distance. All of which proves that there is, indeed, a lot of fact and a lot more fancy swirling on the frontiers of starflight. But never forget that out of this maelstrom of "crazy ideas" may come the *one* we could use to get to the stars—in style!

14 Suspended Animation, Hibernation, and Hypothermia

To die, to sleep;
To sleep: perchance to dream: ay, there's the rub:
For in that sleep of death what dreams may come,
When we have shuffled off this mortal coil,
Must give us pause.

William Shakespeare, *Hamlet,* Act III, Scene 1

The usual thoughts about starflight are with the physical sciences—relativity, astrophysics, the technologies of propulsion, materials, and navigation. But biology and human physiology may yet have as much import for the future of interstellar travel as those "harder" sciences. Whether one is waiting for data to be returned from robot probes or contemplating travel on an interstellar journey, time is of the essence and biology gives one little of it.

Either way, our brief human existence is incompatible with the enterprise of starflight, except to the extent that we are happy to vouchsafe to descendants the joy of future discoveries and experiences from or in other solar systems. We of the twentieth century—the first serious planners and dreamers of starflight—have little chance of being aboard the first peopled starships. It is all but certain that we will never step aboard a worldship, and we have already consigned to our descendants the direct excitement of interstellar voyages.

But nature hints at a solution that will allow the *individual* to make the stars his destination: the hibernating creatures of the forest, who can lower their metabolism and thus survive the barren winter in deep cold. Other clues are borne by the survivors of profound hypothermia, who occasionally emerge unscathed after extended submersion in frigid water or burial in snow. Will the slowing or cessation of metabolism someday allow starnauts to ply the trade routes of the galaxy on long, slow voyages occupying centuries or millennia? Will the drama of suspended animation, played out in numberless science-fiction tales, someday become reality?

Searching for Elysium

First some definitions and priorities. *Biological immortality:* the vanquishing of the effects of aging in an *individual* by genetic/biochemical control. Simple cloning does not count. Amoeba and the like are already immortal, and cloning an "identical" twin does nothing to preserve the identity of mind. *Suspended animation:* the effective cessation of metabolism in a living organism by deep-freezing with the ability to reanimate in the future. *Hibernation:* temporary slowing of metabolism, allowing an individual organism to be immobile for longer than normal periods of time. *Hypothermia:* lowering the temperature of an organism below normal to facilitate hibernation or perhaps suspended animation.

The invention of human biological immortality would create the most serious social crisis short of an apocalyptic thermonuclear war—a paradox of extremes, if ever there was! Yet overnight, biological immortality would make starflight at least theoretically possible for individuals. Undoubtedly a community of immortals could be found who would be willing to make a slow, extended journey to a distant solar system, provided their safety could be ensured absolutely! Radio astronomer Frank Drake has speculated, possibly correctly, that extraterrestrials who had become biologically immortal would be unusually concerned about their physical safety (1). There is little reason to believe that immortal human beings would feel otherwise. So perhaps biological immortality and the inherent risks and uncertainties of starflight would make terrestrial immortals shun a starship's beckoning door. On the other hand, some science-fiction storytellers have suggested immortals who were bored and *craved* risk!

In any event, scientific opinion does not regard biological immortality as being just around the technological corner, but who knows what the coming centuries of biological discovery will bring? Even if we knew the details of the genetic and cellular mechanisms that control aging, and we do not, it is not certain that there would be an easy or even a *possible* molecular fix to suspend aging without literally redesigning our biochemistry. So put aside thoughts of "Elysium" for the time being, and consider things that are already known or that really might be.

Cryobiology

Deep cold is already an extremely useful technique in modern biology. Far below the freezing point of water—0°C—normal atmospheric pressure, molecules cease the frenetic pace that permits chemical reactions to go forward. In frozen silence, the drama of life comes to a standstill for weeks, months, years, perhaps even centuries. Cryobiology comes closer, in effect, to "halting time" than all the inventions of the philosophers and would-be time travelers.

Today cryobiology provides frozen corneas for later transplantation in diseased eyes. Red blood cells in the frozen state may remain intact and reusable a decade or more, whereas unfrozen blood lasts no more than several weeks in cold storage. Several cubic millimeter portions of pancreatic tissue have been frozen, later thawed, and then successfully transplanted into an appropriately tissue-matched animal recipient—one possible route to the cure of diabetes. Small bits of frozen neural tissue have been successfully transplanted into animal and human brains in efforts to control the symptoms of Parkinsonism. And very early stage mouse embryos are now being frozen, thawed, successfully implanted, and brought to term.

Medical researchers routinely preserve solutions of mammalian cells. Artificial insemination of cattle from the frozen semen of champion sires is now a mainstay of advanced agriculture. The applications of deep cold are numerous, indeed, and expanding—a limited version of suspended animation at work today!

The guiding principle of cryobiology: Ice is the enemy of delicate cell membranes and structures. Crystals of ice destroy cells when the freezing of water leads to its physical *expansion*. To circumvent this deadly process, biologists infuse "cryoprotective" agents in tissue before lowering its temperature, such compounds as glycerol and dimethyl-sulfoxide. The protective agents surround the cells and replace the hazardous extracellular water, which constitutes an average of 80% of cellular volume. Below the temperature $-130°C$, the so-called *glass transition temperature*, crystals of water-ice can not form. The trick is to drop the temperature at an appropriate optimal *rate* for a particular kind of cell, for instance, a few hundred °C per minute for red blood cells protected with glycerol. In rewarming, optimal temperature rates of elevation must also be respected.

But a few cubic millimeters of tissue are far different from whole organs, much less complete organ systems—whole bodies. The problem with these larger masses of cells is that they are highly differentiated, and many different cell types coexist. Each cell variety requires a different freeze-thaw rate profile, an inherently impossible requirement that is lessened somewhat by cryoprotective agents. But how to completely treat a large organized mass of cells with these agents? The permeation of the tissue mass with cryoprotective agent takes a long time and may itself cause harm through various kinds of chemical osmotic gradients. At the moment, the freezing and subsequent successful thawing of a large organ mass is impossible and may always remain so, some cryobiologists suggest.

This conclusion of modern science is blithely ignored by contemporary "cryonicists," those seekers of immortality who are exponents of freezing "cryoprotected" deceased humans. They harbor the surely vain hope that the all-knowing clinicians of the far future will be willing and able to thaw and restore them and their brain-based human identities. In fact, some cryonic entombments have adopted the economic expedient of freezing only the severed head, presuming that future resurrectors will provide it with a matching body! It is, therefore, understandable that the bylaws of the entirely legitimate Maryland-based international scientific organization, The Society for Cryobiology, Inc., spell out a prime reason for expulsion from the society: ". . . misrepresenting the science of cryobiology, including any practice or application of freezing deceased persons in the anticipation of their reanimation."

Hypothermia and Hibernation

Yet, if anecdotal evidence published in the conservative medical literature is to be believed, what may be highly questionable practice today may be the norm of the future. Indeed, the chilled "apparently dead" seem in many instances to have risen. The reason this can be is that the brain tolerates loss of oxygen many times better at low temperatures than at normal body temperature of 37°C.

John Hands has reviewed some startling cases in which physicians managed to revive profoundly hypothermic patients (2). Submerged in chilled water for up to an hour or buried in an avalanche of snow and ice for up to five hours, these victims were brought back to life without perceptible neurological damage. This, after complete cardiac arrest—often for hours—and core body temperatures falling to near 20°C. The medical procedures were various but generally involved surgical opening of body cavities and the warming of internal organs with heated fluids. In fact, surgeons have used the lessons of these accidental hypothermic experiences to perform successful open heart surgery by surface body cooling, rather than by oxygenating the blood and replacing it at normal temperatures.

The rules of emergency medicine have changed and a new one has been added: "No one who is not both warm and dead should be considered dead." Now it is still a long step from deep hypothermia for a transient period to suspended animation for a long time. But may we not expect scientific surprises akin to the one received by physical scientists in recent years working on superconductivity? Few physicists thought that superconductivity—the absolute vanishing of resistance to the flow of electrons—would manifest itself readily much above 20°K. Yet since 1986 the completely new phenomenon of high-temperature superconductivity in special ceramic materials has arisen (at this writing, above 100°K), and researchers seriously project room temperature or higher superconductivity. Biologists be alert and aware!

One possible road to life extension in deep cold is to explore the possible connection between hypothermia and the natural hibernation of animals. The two phenomena are superficially similar, but physiologists who have studied their comparative aspects conclude that there are also major differences (6). In both processes, for example, there may be lowered oxygen uptake and reduced respiration and heart rates. But the mechanisms of onset and emergence from hypothermia and hibernation are clearly different. Entry into hibernation involves a sequence of physiological and behavioral stages that may never reach a continuous state, "bouts" of days or weeks of hibernation being the common mode. Nevertheless, body temperature is lowered in the hibernating state from normal temperatures in the vicinity of 38°C to the range 4 to 7°C.

The most fundamental difference, of course, is that hibernating animals spontaneously emerge from their torpor, whereas hypothermic animals do not. Hibernation is the product of evolution, while induced hypothermia is an externally imposed condition for which the body typically has inadequate defenses. Induced hypothermia experiments with normal hibernators, for example, hamsters, reveal that metabolic regulation in hypothermic subjects is insufficient for extended life preservation. Experiments reported by Mussacchia and others suggest that infusing glucose during hypothermia enhances survival by correcting the hypoglycemic condition (5). There are many metabolic distinctions between hibernation and hypothermia involving not only blood glucose levels but also oxygen demands, acid-base balance, and central nervous system biochemistry—too intricate and varied in different animals to review here.

The mechanism of death in profound hypothermia in humans is not cell damage by ice crystallization but invariably ventricular fibrillation of the heart. Experiments with hamsters, on the other hand, point to respiratory failure as the proximate cause of death in hypothermia. All of which shows the difficulties blocking the experimental path to ameliorating the effects of hypothermia, perhaps by mimicking some of the metabolic regulation found in hibernation. Hyperbaric (high-pressure) oxygenation, glucose infusion, electrolyte regulation, heat input, and other external controls are some of the ways that an artificial "hibernation"—a long-duration and reversible hypothermia—might conceivably be induced. It is by no means certain that this could be brought about in human subjects and with what long-term effects, but we surely lack sufficient knowledge to prove that it can not be done.

The Dream of Suspended Animation

Suspended animation is nothing new. It is routine today that frozen semen, ova, and even embryos are thawed and then injected or transplanted to begin new life, with the considerable assistance, of course, of warm humans or other animals. The growth of fertilized ova or thawed embryos could, in theory, proceed in an artificial womb—*in vitro*, literally "in glass," as the biologists say. If only one knew how to muster the proper nutrients and growth factors. Easy to say, but to carry out the project is probably far more difficult than one now imagines. Yet the level of technological belief in such a prospect is strong. Growth of fertilized ova or thawed embryos *in vitro* will be done before long, perhaps first with mice and inevitably humans.

There will be people born of no mother. Perhaps it will be they who will emerge from embryonic frozen silence and be the first human emissaries to the stars. But how would they develop, how be raised without the care and love of parents or surrogates? No easy answers, to be sure, and perhaps this unappealing approach to peopled ("embryoed"?) interstellar missions may be almost literally a dead end.

As long as the hardware of starflight remains in the embryonic planning stages, people will dream of the short cut: true suspended animation of human bodies and revival at destination. We know far too little either to dismiss the prospect entirely or to embrace it as a panacea, but the concept is surely too valuable to throw out. As with the hardware of starships, we should remain open-minded about three major biological prospects, dreams worth having: (1) long-duration human hibernation; (2) bio-engineered life-extension at normal temperatures; and (3) indefinite suspended animation.

15 Scientific Payloads

Once acclimatized to space living, it is unlikely that man will stop until he has roamed over and colonized most of the sidereal universe, or that even this will be the end. Man will not ultimately be content to be parasitic on the stars but will invade them and organize them for his own purposes.

J. D. Bernal, *The World, the Flesh, and the Devil*, 1929

Twinkle, twinkle, little star,
how I wonder what you are.
Up above the world so high,
Like a diamond in the sky.

Anonymous

Scientific payloads are the *raison d'être* of starflight. We are not too concerned with the kinds of scientific gear that an interstellar *colony* would take along. Suffice it to say that if 1000 folk are going to live in a "can" for 1 or 40 generations, they will have brought with them every piece of scientific equipment known to humanity, so that at journey's end they will have no trouble exploring their new solar system inside out. Perhaps they will even have developed some new instruments along the way. The interstellar colonists will require plenty of space aboard their efficient ark for all these goodies and their means of delivery—land rovers, mountain-climbing robots, research submarines, aircraft, balloons, remote observation satellites, and so on.

What one should really be concerned about now, however, are payloads for automated probes, presumably among the first craft to depart the Solar System. Mass and space will be extremely limited, yet scientific return must be maximized, so how massive do interstellar scientific payloads have to be? The answer depends so strongly on the level of technology at the time of launch that it is hard to hazard a guess. But to begin to judge the difficulty, imagine early twentieth-century scientists speculating about the mass required for an instrumented interstellar probe! Apart from their complete ignorance of advanced digital microwave signaling, miniaturized computers, and sensitive radio receivers, they would have been at a loss to conceive of the kinds of remote-sensing instruments now at our disposal.

Our blind spot in the late twentieth century—more so than our skepticism about high-velocity starflight—might well be the degree of possible microminiaturization of instruments and computers for a star-probe. Robert L. Forward's bold mid-1970s suggestion that an array of intelligent and redundant microscopic circuitry could coat the few grams of wires of his cobweblike Starwisp probe might be closer to the abilities of future technology than we imagine. We are inclined to believe that it will eventually be possible to build these extremely low-mass interstellar probes—*microprobes*—in the few grams or subgram mass range. These probes will have to rely on observations and data transmission made by phased-array or interferometric methods, and at destination they will have to capture power from the new sun. Of course, the lower the mass of the probe, the more feasible high-speed star travel becomes.

Up to this moment we have talked as though the payload was to be designed around the propulsion system, but it may turn out to be the other way around. If, for example, payloads on the order of only grams really become possible (given a large energy collecting sail-sensor-antenna combination), then beamed power propulsion or solar sailing would be much more attractive than any kind of rocketry. It is easy to see that for a Daedalus-class nuclear pulse vehicle, a few grams of payload would be absurd. So much engine structure remains when Daedalus' fusion fuel burns out that 400-plus tons of payload is a reasonable additional increment. Yet 400 tons seem extraordinarily luxurious for a probe that merely will fly through an extrasolar planetary system, even allowing for subprobes deployed for individual planets at terminus. On the other hand, if any extensive planetary landings, biological sampling, and surface roving are contemplated, then microprobes would be inherently less feasible. For those ends, heavy, rugged, and environmentally protected vehicles, decelerators, and escape rockets would clearly be necessary.

101 Things to Do Along the Way

Sweet are the uses of adversity: Along with a lot of boring time to "kill," there are a multitude of experiments that can be conducted on the long journey to a star. Since there is uncertainty about the configuration and morphology of instruments that would be carried on an automated interstellar probe, it is appropriate at present to characterize them only according to function and objective. In summary:

• Measure *in situ* the density, charge, mass, species, velocity, and temperature characteristics of interstellar plasma and gas.

• While on the initial leg of the interstellar trajectory, define the location and plasma properties of the heliopause between the solar wind and the interstellar medium. Approaching the destination solar system, determine the location and characteristics of its heliopause.

• Looking back at the Solar System, test autonomous planet/moon detection instruments. Observe the Solar System "as a whole" in various radiation bands to measure its gross properties and "to see ourselves as others see us."

• Define the conditions of the medium in the Oort comet cloud as well as the extent of the cloud, for instance, detect comet nuclei with radar. Determine comet frequency distribution as a function of radial distance from the Sun.

• Continuously measure the orthogonal components of the galactic magnetic field.

• Continuously monitor the interstellar medium for molecular species and micrograins. Determine mass, composition, size distribution, and frequency of interstellar grains. Perform interstellar erosion experiments with various models of shield configurations.

• Use the long-baseline formed by the starship and Solar System to carry out high-resolution astronomical measurements with optical and radio interferometry. Perform astrometric measurements of nearby stars, extrasolar planet detection, extrasolar planet imaging, and atmospheric spectroscopy. Use the same long-baseline techniques for astrophysical measurements, for example, image radio-galaxies, quasars, and neutron stars.

If multiple starships are underway at the same time, the capability of this long baseline interferometry could be enhanced by combining their observations of distant sources. A new difficulty in this kind of interstellar interferometry: common time-tagging of data from probes moving in different directions at significant fractions of the speed of light.

• Observe low energy cosmic rays (normally excluded from the confines of the Solar System).

• By monitoring the Solar System escape trajectory through careful radio tracking, determine the possible existence of an undiscovered Solar System stellar companion; discover possible "brown dwarfs" along the way through gravitational perturbations of the starship trajectory.

• Monitor the interstellar medium's effect on the starship's optical and radio transmissions back to the Solar System. Observe scintillation due to zones of varying interstellar plasma density.

• Try to detect gravity waves from astrophysical processes (supernovae, neutron stars, etc.) by monitoring anomalous Doppler shifts in starship signals.

High-velocity interstellar flight presents difficulties for some kinds of measurements, particularly those aimed at determining the nature of interstellar grains. It would be desirable to capture interstellar grains "unscathed" because their examination by automated microscopy could reveal aspects of how they were formed that would be impossible to know with destructive capture. The virtues of examining everything thoroughly along an interstellar path are apparent, but it is less clear how to accomplish this short of intolerably slow flight. Perhaps intense magnetic or electrical fields could be used to decelerate small grains and capture them.

Planetology

Even though a large energy cost must be paid to decelerate into another solar system after a high-speed interstellar voyage, we are partial to such missions. We believe that a rapid fly-through, such as envisioned in Daedalus, would be inherently orders of magnitude less informative than a mission that permitted detailed planet examination by orbiters, entry probes, and landing spacecraft.

Any experiments that have been done or contemplated for Solar-System planets are candidates for orbiting/landing missions of extrasolar planets and moons. One major difference, however, is that the identity and character of the planets within the target system may be relatively unknown. Autonomous astronomical instrumentation will be required to locate and determine the orbital elements of bodies within the new solar system. Once they are found and probes are dispatched to them, the science objectives are obvious:

- Synoptic imaging and mapping from orbit in visible, infrared, and ultraviolet wavelengths

- Surface imaging from roving landers, balloons, and long-duration aircraft

- Radar mapping from orbit whether or not a planet is shrouded by clouds

- Magnetic field and magnetosphere plasma measurements

- Measure interplanetary dust

- Atmospheric entry probes to determine atmospheric composition

- Determine ocean composition and extent

- Seismological monitoring

Absent human observers, a high degree of computer intelligence will be required to coordinate and analyze the influx of scientific data to determine how the exploration should unfold. Reexamining the case history of how Solar-System planets were investigated in the twentieth and twenty-first centuries will be critical to providing the starship computer with the relevant expertise.

In our enthusiasm to explore extrasolar planets, let us not forget that this will be an exceptional opportunity to study another star (possibly with a companion sun) close up. The scientific package should contain instruments to analyze the "stellar wind" and magnetic fields, image the star's surface in all possible wavelengths, observe the star's rotation, monitor flares, and study the interaction of stellar components.

Exobiology

The automated search for extraterrestrial life within another planetary system will be among the most exciting and challenging aspects of the scientific mission of a robotic explorer. All such searches will be hampered by the difficulty (or perhaps impossibility) of defining universal properties of living systems. But a starting point will be the the kind of instrumentation that accompanied the Viking landers to the Martian surface in 1976 (Figure 15.1). The Viking life-detection experiments were designed to check for metabolic activity in potential Martian organisms cultured from soil. Unfortunately, the result of those experiments was ambiguity sufficient to cause some researchers to claim that Viking did, indeed, uncover evidence of life. With that as background, future experiments

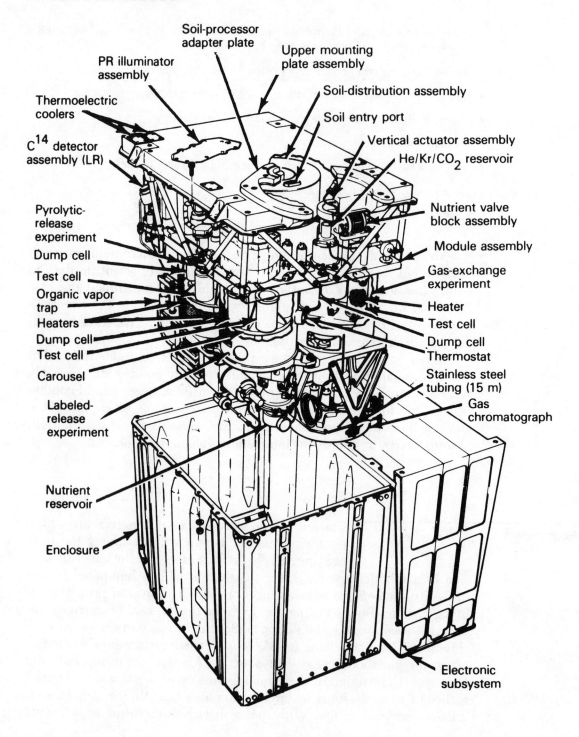

Soil-processor
adapter plate

Upper mounting
plate assembly

PR illuminator
assembly

Soil-distribution assembly

Thermoelectric
coolers

Soil entry port

C^{14} detector
assembly (LR)

Vertical actuator assembly

He/Kr/CO_2 reservoir

Pyrolytic-
release
experiment

Nutrient valve
block assembly

Dump cell

Module assembly

Test cell

Gas-exchange
experiment

Organic vapor
trap

Heater

Heaters

Test cell

Dump cell

Dump cell

Test cell

Thermostat

Carousel

Stainless steel
tubing (15 m)

Labeled-
release
experiment

Gas
chromatograph

Nutrient
reservoir

Enclosure

Electronic
subsystem

Figure 15.1 Viking life-detection apparatus. The 1 cubic-foot automated biological laboratory was carried to Mars by each Viking spacecraft. (Courtesy NASA)

will be designed to rule out the kind of inorganic chemical activity that has been used to explain the Viking results.

The morphology and dynamical behavior of living systems may be the biggest clue to the existence of alien life forms, so it may be that imaging a planet surface at various length scales—including the microscopic—may be the most appropriate technology for searching for life. Whether computer programs can be developed to scan such imagery for life forms remains to be answered. That may be a key question because we cannot rule out the possibility that a landing probe could be damaged or destroyed by indigenous species, unless it could take steps to evade inadvertent threats by alien creatures. There is no need to even speculate about what to do if the critters (or plants) have spears—we should be so fortunate to find such a world!

Communication and Reliability

Communication of the scientific data from an interstellar probe will involve three major subsystems: (1) the main downlink to the Solar System by an antenna/transmitter system; (2) the interplanetary communications network of probes, relay satellites, and receiving/tracking antenna(s) on the mother ship; and (3) the Solar System-based antenna/receiver system. It should be assumed that communication to the Solar System will be at microwave frequencies, thus taking advantage of Earth's existing radio astronomy antenna infrastructure—the VLBI network, the Arecibo antenna, and possibly large aperture antenna systems (e.g. the once proposed Project Cyclops array) designed for SETI "leakage signal" listening efforts. It is possible, however, that laser transmission of intersolar system data will be deemed more efficient and appropriate.

The precursor of an interstellar data communication system will be exercised in August 1989 for the Voyager-2 encounter with planet Neptune, about 30 AU from Earth. A single electronic image of Neptune will consist of about 5 million data bits (800×800 pixels, 8 bits/pixel), but this will be reduced before transmission 60–70% by advanced data compression techniques. During the Neptune encounter, antennas on Earth will be required to capture data at rates up to 21,600 bits/sec, a feat that will be accomplished by the synthesis of a large antenna aperture consisting of the 27 dish Very Large Array (VLA) near Socorro, New Mexico, and the Goldstone Deep Space Communications Complex in Goldstone, California (two or three antennas). Yet the antennas will be gathering only a minute fraction—10^{-17} watts—of the mere 22 watts of transmitter power aboard Voyager-2.

Bit rates required to make the return of scientific data from nearby stars useful, surprisingly, may be much less than for Voyager at Neptune encounter. For one, significant advantages may be gained by data recording and playback over long-time intervals (years as opposed to minutes). There is certainly no need for "real time" transmission of imagery from extrasolar planets. To sustain higher transmission rates may involve transmission at higher power, a higher gain transmitter/ antenna, and a larger effective aperture receiving network—or a combination of all three. The designers of the Daedalus communications subsystem chose to do it in style and selected a microwave radio link that would broadcast 864 kilobits/sec at a power output of one megawatt from Barnard's Star (5.9 ly). It appears that communicating data across interstellar distances is well within our current technological capacity. Moreover, our ability to sustain higher transmission rates is likely to escalate in the coming decades, if past progress is any guide.

Needless to say, the reliability of the scientific instruments as well as the communications subsystem will be of paramount importance. After traveling so long and at great cost, catastrophic failures in these systems would be intolerable. The problem is especially acute because the systems will be exposed to the rigors of interstellar space for decades, if not centuries. Radiation shielding, radiation hardening, and redundant components will be essential throughout. It is necessary that the computer architecture for an interstellar mission be fault-tolerant to an extreme: Failures in one area could not be allowed to propagate destructively to other areas. The technology of such fault-tolerant computers is now developing nicely, the byproduct of aerospace missions of varying kinds, deep space missions in particular.

Encountering another planetary system and exploring it in great detail will be virtually a recapitulation of the history of astronomy and early space exploration, except that there will be new paths to follow and new surprises along the way. But the new discoveries will not be made directly by human astronomers, but by reliable and savvy robotic emissaries whose prodigious intelligence may be the first to perceive differences and similarities with bodies of the Solar System. These automated craft will be the pathfinders for the human explorers and colonists who will surely follow.

16 Detecting Extrasolar Planets

For the universe is infinite and therefore without centre or limit . . . especially when we demonstrated that there are certain determined definite centres, namely the suns, fiery bodies around which revolve all planets, earths, and waters, even as we see the seven wandering planets take their course around our sun.

Giordano Bruno, *On the Infinite Universe and Worlds,* 1584

And thus we die,
Still searching, like poor old astronomers
Who totter off to bed and go to sleep
To dream of untriangulated stars.

Edward Arlington Robinson,
1869–1935,*Octaves*

Giordano Bruno, gagged and burned alive at the stake in Rome on February 17, 1600, went to his horrible death for believing in planets of other stars and an infinity of inhabited worlds. Almost four centuries later, the average person, if asked, is likely to know that Earth and other planets orbit the Sun. Sadly, that same citizen in one of the world's most advanced technological civilizations is less likely to know that those points of light—the stars—are merely distant suns, many if not most of which are likely to have planets. How slowly we learn!

A Venerable Problem

If we could prove without leaving the Solar System that *no* planets attended other suns, interstellar travel would still be an interesting engineering challenge, but its objective would then be so much less compelling. Even though there is yet no *definite* and *unambiguous* proof of the existence of extrasolar planets, particularly small, Earth-size ones, the case for such worlds has been overwhelmingly established through indirect evidence. Astronomers are within a few years—a decade at most—of having conclusive evidence.

A starflight compendium would certainly be incomplete without telling of the great quest to detect planets circling other stars. Even with ground-based techniques, the field is now approaching a crescendo of activity, but it is about to grow explosively as large spaceborne instruments, such as the Hubble Space Telescope (HST), become operational.

Until the mid to late 1940s, many astronomers would have considered any extrasolar planet search to be futile, no matter how advanced its technology. The prevailing planet formation theory of the time—the so-called "tidal theory"—required an unlikely chance or grazing encounter between two stars to form planets. A gaseous filament gravitationally pulled out of one or both of the stars in the rare near-collision was the supposed source of condensed, spinning worlds.

If the tidal theory were correct, very few or no other planetary systems would exist in the galaxy because grazing stellar encounters are extremely infrequent. The overwhelming weight of astronomical evidence and virtually all astronomers today support the more venerable resurrected "nebular hypothesis": All Solar-System objects—the Sun, planets, moons, asteroids, and comets—condensed approximately 5 billion years ago from an interstellar nebula or cloud of dust and gas. Formation of young stellar objects—planetary systems included—has been occurring for billions of years in nebulae such as M42 in Orion. By some accounts, more than 50% of all stars have attendant planetary systems. So starship destinations will not be uncommon in the galaxy, and on suitable worlds extraterrestrial life may also be plentiful.

Planets can be divided into three categories, forming perhaps a continuum of masses of secondary bodies circling a primary sun (or one or both suns in a binary star system). Two of these classes are represented by worlds circling Sol: (1) the "gas giants" or Jovian worlds, enormous low-density planets with extensive atmospheres, many satellites, and ring systems like Jupiter, Saturn, Uranus, and Neptune and (2) the smaller "terrestrial" planets, such as Mercury, Venus, Earth, and Mars (Pluto is a bizarre, still difficult-to-define anomaly) having few or no satellites, comparatively thin atmospheres, and higher densities than the Jovian worlds. Jupiter, the largest and most massive world in the Solar

System, is about 300 times the mass of Earth but only 1/1000 the mass of the Sun. A third category of planetlike objects, intermediate in mass between gas giant worlds and small stars, are called "brown dwarfs" with masses up to 85 Jupiter equivalents and relatively cool temperatures (120 to 2000°K). (The term brown dwarf was apparently coined by astronomer Jill Tarter, who has been prominent in SETI research.)

Before charging off to problems of planet detection, consider what types of worlds we might reasonably hope to detect across interstellar distances. For one, technology that would allow astronomers to find "rogue" planets—hypothetical free-wandering interstellar worlds (brown dwarfs or smaller)—is difficult to envision, though perhaps observing the perturbations of starship trajectories will be of some use. The best prospect for the immediate future is to confirm the presence of Jovian-size planets near other suns. Smaller denizens of another planetary system—small rocky planets, their moons, asteroids, and comets—will be much harder but *theoretically possible* to detect and even form images of at interstellar distances.

Only in the last few years has technology advanced to levels sufficient for the potential detection of gas giants or brown dwarfs. It is generally accepted that many of the early claimed discoveries of extrasolar planets have turned out to be false, though in no way does this rule out planets around those same stars. A truly confirmed detection will require independent observations, using at least two different instruments or measuring techniques.

Extrasolar planet detection is difficult for three fundamental reasons: (1) the brightness ratio of a star's self-generated illumination to reflected radiation from any of its planets is huge, depending on the wavelength band, factors of thousands or more often, billions; the brightness ratio is much more favorable (smaller) at infrared wavelengths than in the visible; (2) the angular separation of a planet and star at interstellar distances is also extremely small, from tiny fractions of an *arc second* (One arc second is 1/3600 of a degree of angle—about the apparent diameter of a golf ball seen at a range of about 5 km.) to several arc seconds at best; and (3) the great disparity in planet versus stellar masses, a fact that makes some indirect detection methods more difficult.

Astronomer Jill Tarter and her colleagues have enumerated four generic approaches to detecting extrasolar planets: (1) direct detection, that is, in some sense directly imaging the planet; (2) *astrometry* or detecting periodic changes in the parent star's position caused by the orbiting planet; (3) *spectroscopic detection* of stellar velocity shifts due to planets; and (4) *photometric detection* of variations in stellar luminosity caused by eclipses and related phenomena (2). The latter three methods are, of course, methods of indirect detection.

Direct Detection

Early work on direct extrasolar planet detection, performed between 1959 and 1977, assumed that these worlds would first be observed using space telescopes rather than terrestrial instruments. These proposals embodied two basic approaches: single telescope detection and interferometry.

Single mirror telescopes—such as a space telescope—were first investigated by Nancy Roman in 1959 and Lyman Spitzer in 1962 (3,4). Early treatments suggested placing an occulting disk between the telescope and the primary star. The occulter was to reduce light from the primary so that the much dimmer orbiting planet would become visible (Figure 16.1). Alphonsus Fennelly and colleagues expanded in 1975 on an earlier suggestion by Nancy Roman that the Moon's limb could serve as a

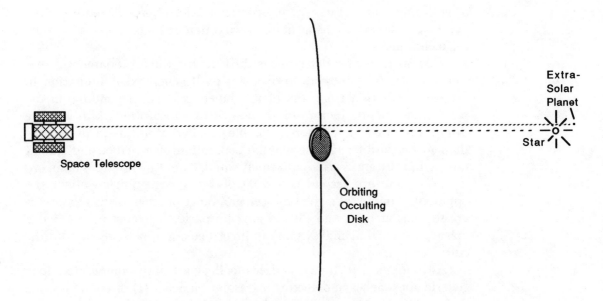

Figure 16.1 Space telescope occulter for extra-solar planet detection.

suitable occulter (5). James Elliot of MIT, then at Cornell, further developed the lunar occultation approach (6).

In 1968, Gerard K. O'Neill of Princeton suggested that an orbiting optical interferometer—or multiple mirror telescope—could image extrasolar planets and not require an occulter (7). Optical interferometry in space, though difficult and expensive, turns out to be one of the most promising methods of planet detection, as we shall see. Much of the early work on both direct space observation approaches has been reviewed by Huang, Martin, Matloff and Fennelly, and Kenknight (8–11). More recent work has been summarized by Baum, Black, Russell, Tarter, and others. As described by Tarter and her associates, the more recent reviews are more accurate in their treatment of the spatial resolution of single-mirror telescopes used as planet detectors (Technical Note 16–1).

Earlier studies were far from sanguine on the possibility of detecting planets from the Earth's surface. They regarded poor astronomical

Technical
Note
16–1

Direct Detection

As described by Tarter et al, the brightness ratio of the extrasolar planet to its parent star, H, is a key parameter for planet imaging (2). For Jupiter and the Sun, $H = 2 \times 10^{-9}$ in the visible, 10^{-4} in the infrared, and 3×10^{-3} for millimeter radio frequencies. For the Earth and Sun, $H = 10$ to 100 times smaller. In addition to higher H than the visual region, another advantage of longer wavelengths is the fact that much of Jupiter's flux is produced internally from gravitational energy.

The correct expression for the smallest resolution, $\Delta\theta$, in the case of a planet much dimmer than its primary star, is:

$$\Delta\theta = \frac{7.11 \times 10^4 \lambda}{H^{1/3} D_0} \text{ arc seconds}$$

where λ = wavelength and D_0 = telescope diameter.

In imaging the planets of our Solar System from a nearby star, one would observe as Matloff and Fennelly noted(5), that the Earth's reflected sunlight would be more blue-shifted than reflected light from the other planets. Color shifts of planet images from the primary star's color could be helpful in planet classification if combined with information regarding the separation of planets and primary star.

"seeing" within the atmosphere as an insurmountable difficulty. Normally, the optical resolution of a small or large aperture telescope looking through the atmosphere is limited to about one arc second or slightly better, a resolution barrier imposed by the inhomogeneous and dynamic atmosphere. In recent years, a mathematical image reconstruction technique to improve seeing—*speckle interferometry* (invented by French astrophysicist Antoine Labeyrie)—has broken that barrier and made ground-based planet-detecting schemes theoretically feasible.

Thousands of individual image frames are recorded (at a rate of about 100 per second) using a sensitive electronic camera incorporating a solid-state *charge coupled device* (CCD) at the telescope focus. Each frame contains sky data in which point sources are blurred by atmospheric turbulence and telescope imperfections. By appropriate mathematical processing, a significant amount of atmospheric interference can be removed and resolution improved. But even in space and using an ideal telescope/recording system, the optical image of an extrasolar planet will appear as a buried "signal" in the much larger "noise" of light from the primary star's inevitable spread-out diffraction pattern of light. As initially pointed out during a 1976 Lick Observatory workshop chaired by Jesse Greenstein and more recently reiterated by Robert Brown of the Space Telescope Science Institute, direct imaging of an extrasolar planet is further complicated by slight imperfections in the optical figure of even the best telescope (14,15). Using speckle interferometry in the more favorable infrared spectrum, some researchers have managed to detect dim companion stars, a step toward detecting extrasolar planets by similar means.

Putting these problems aside for the moment, it is clear that attempts to image extrasolar planets will be made using the Hubble Space Telescope when the Earth-orbiting observatory finally becomes operational, it is hoped, in 1989. As described by J.L. Russell of the Space Telescope Science Institute, a number of science instruments aboard the HST could be applied in planet searches. These include the Faint-Object Camera (FOC), the Wide-Field Camera, and the High-Speed Photometer. The FOC is equipped with a "coronagraph finger" that could operate as an internal occulter to reduce light from the primary star to make imaging of Jovian worlds, at least, a possibility. NASA also intends to deploy a number of infrared telescopes in space during the 1990s, including SIRTF, a one-meter diameter telescope planned to be 1000 times as sensitive as the pioneering IRAS satellite (see below). These could also serve in the search for Jupiter-sized planets of nearby stars.

Indirect Detection Methods

Astrometric Methods

Unlike direct imaging, indirect techniques do not aim to return an image of an extrasolar planet, pointlike though that image may be. Rather, in indirect methods the planet reveals itself through circumstantial evidence. In the classic indirect approach—astrometry—researchers have used long focal-length refracting telescopes in planet searches. A Jupiter-size planet orbiting in a celestial dance with its primary star around the pair's *barycenter* (center of mass) alters the position of the parent star—as viewed against a background of more distant reference stars—in a periodic manner. If the relatively nearby primary star also has a relatively high proper motion and is less massive than the Sun, the "wobble" introduced to the star's otherwise straight-line path is theoretically detectable (Figure 16.2).

In an astrometric planet search, decades of photographs of the subject star must be acquired to capture enough of the cycle of variation caused by the supposed planet(s). Variations in the star's motion are then analyzed by comparing the star's and more distant "stationary" reference stars' images on photographic plates with an accurate "measuring engine." After being obtained by the measuring engine, the star image data is used as input to a position error-removing computer program. Recent progress in astrometric detection is reviewed by Tarter and others (2). Peter van de Kamp, former director of Sproul Observatory at Swarthmore College, has published an excellent reference source in the science of photographic astrometry: stellar position measurement (28).

Unfortunately, astrometric planet search results for the same star observed using different telescopes have not generally been in agreement,

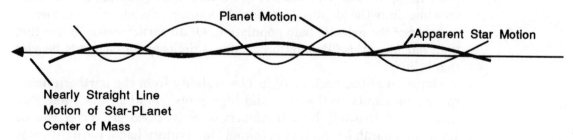

Figure 16.2   A planet's perturbation of a stellar path.

perhaps because of systematic instrument errors. Some long-focus astrometric refractors, such as the one at Allegheny Observatory near Pittsburgh, are being upgraded for more advanced searches (19,20). Space-based astrometric equipment is also being considered. Even with these improvements, however, astrometric planet searches will be uncomfortably close to fundamental limitations. We cannot expect that planets much smaller than Jupiter or Saturn will be detectable using astrometry, even if they orbit the nearest stars. Nonetheless, a great advantage to astrometric planet searches is the vast amount of photographic imagery of nearby stars in the archives of the dozen or so university observatories that have been involved in astrometry for decades. With the advent of the Hubble Space Telescope, its Wide-Field Camera will be brought to bear in the first astrometric planet searches from space.

The most famous controversy in astrometric planet-detection is that surrounding the presumed planetary system of Barnard's Star. Although tantalizing, claims of detected planets depend on data very close to the accuracy limits of present-day observational techniques. Discussed in recent reviews by Ken Croswell of Harvard University and Robert and Betty Harrington, Barnard's Star is one of the most studied stars in the sky, and also the one with the highest proper motion (21). A red dwarf with 15% of the Sun's mass and 1/2000 of its luminosity, Barnard's Star was discovered in 1916. It is the second nearest star system and the fourth nearest star to the Sun at a distance of 5.95 ly—only Alpha Centauri A and B and Proxima Centauri are nearer.

Barnard's Star is moving across the northern skies in the vicinity of the constellation Ophiuchus at 90 km/sec. Its apparent position changes by the Moon's angular diameter in only 175 years. In addition, it is moving toward the Sun at 108 km/sec, giving it a healthy total velocity (relative to the Sun) of 140 km/sec. Because of its high proper motion, Barnard's Star is not a permanent member of the solar neighborhood. After it approaches within 3.8 ly of the Sun in A.D. 11,800, it will begin receding into the depths of space. It appears that it is a wandering member of the galactic halo population. Of an earlier stellar generation than our Sun, this "Population 2" star is deficient in elements heavier that helium.

Barnard's Star, because of its observability from the northern hemisphere, proximity to the Sun, and high proper motion, has been photographed extensively by astronomers since its discovery. Peter van de Kamp and Sarah Lippincott at Sproul Observatory had a Barnard's Star data base consisting of thousands of photographic plates taken between

1918 and 1962. In 1962, van de Kamp announced that analysis of this enormous data set revealed that Barnard's Star was attended by a planetary system. The results of the initial analysis led him to believe that Barnard's Star was attended by a planet with 1.6 times Jupiter's mass, orbiting the star once every 24 years in a highly elliptical orbit. A 1969 recalculation by van de Kamp indicated that *two* planets intermediate in mass between Jupiter and Saturn and in circular orbits was a better fit to the observational data.

Using different telescopes, measuring engines, and data reduction techniques, a number of astronomers have attempted to confirm van de Kamp's discovery. Possible systematic errors in the Sproul data set have complicated the process, as has the fact that no other observatory has a Barnard's Star data set as extensive as Sproul's. George Gatewood and Heinrich Eichhorn of the Allegheny Observatory, working with a few hundred plates from the Allegheny and Van Vleck Observatories, report no perturbation to Barnard's Star's straight-line motion. However, their data set contains consecutive photographs for only a fraction of the period that Barnard's Star has been observed at Sproul. Now Gatewood and his colleagues, having developed an instrument of superlative precision (the computerized MAP or Multi-Channel Astrometric Photometer), have embarked on a much more sensitive astrometric survey of nearby stars, including Barnard's. And it is all electronic, no more photographic plates!

Robert Harrington of the U.S. Naval Observatory has used the 61-inch reflector in Flagstaff, Arizona, to photograph Barnard's Star hundreds of times since 1972. Harrington's data, in concurrence with measurements at the Leander McCormick Observatory in Virginia reported by Laurence Fredrick, show weak evidence for a possible perturbation of Barnard's star's path. It is too early to know whether this evidence will lead to confirmation of a Barnard's Star planetary system. Even if such a system exists, it may be of little use to human colonists because of the heavy-element deficiency in "Population 2" stars and the low luminosity of Barnard's Star.

Spectroscopic and Photometric Detection

Spectroscopic planet-detection schemes, like astrometric methods, rely on the gravitational interaction between planet and parent star, which produces orbital motion about their common center of mass (Technical Note 16–2). If the planet's orbital plane is not exactly perpendicular to

Spectroscopic Detection

As discussed by Tarter et al, a star's spectral lines will be slightly blue-shifted by an extrasolar planet when the star-planet barycenter is between the Sun and the star's center and red shifted when the barycenter is behind the star's center, as viewed from the Sun (2). The planet orbiting the star causes it to alternately approach and recede from the Sun.

For a planet orbiting its primary once every year, in an orbit with a 90° inclination to the tangent plane to the celestial sphere, the stellar velocity corresponding to the wavelength shift can be written:

$$\Delta V \approx 30 \frac{M_p}{M_{Jup}} \left[\frac{M_0}{M_*} \right]^{2/3} \text{ meters/second}$$

where M_p, M_{Jup}, M_0, and M_* respectively denote the masses of the planet, Jupiter, the Sun, and the star.

If an extraterrestrial civilization were looking at the Sun spectroscopically in the plane of the Solar System ecliptic, it would notice about a 12 meter/sec maximum radial motion induced by the Sun's dance with Jupiter around their common barycenter.

the line of sight from the Earth, the Doppler effect periodically alters the frequencies (colors) of absorption and emission lines in the primary star's spectrum. Very high spectroscopic accuracy (a few parts in 10^8) is required. Planetary orbits and sizes cannot be determined and a dim secondary star could be confused with a spectroscopically detected planet. Periodic motions in the stellar atmosphere could conceivably mimic planet-caused motions, and thus the technique requires further development.

On the positive side, fairly modest Earth-bound equipment can be used in a spectroscopic search and the technique is not strongly dependent on the star's distance. As reported by G. Flint, at least one *amateur* astronomer is conducting a spectroscopic extrasolar planet search (16). Professional astronomers, such as the late K. Serkowski, have also conducted spectroscopic searches (17).

Photometric planet detection schemes depend on the fortuitous alignment of the extrasolar planet's orbital path around its primary star, as viewed from the Earth. One out of 20 Mercury-size planets and one out of every million Jovian worlds will periodically partially occult or eclipse the disk of its primary star, as viewed from the Earth. As W. J. Borucki and colleagues have demonstrated, the duration of such an eclipse will be days, and the stellar flux might be reduced by 1% during an occultation by a "Jupiter" and 0.01% by an "Earth" (18). Photometric precisions of 10^{-5} are necessary to detect the luminosity changes and 2×10^{-6} to monitor color variations. Observations should be made during the quieter phases of the star's activity cycle.

Optical Interferometry

Optical interferometry based in space or on the surface of the Moon beyond the year 2000 may allow the imaging of extrasolar planets at fantastically high resolutions—sufficient to see terrestrial-type planets in continent level detail. According to Professor Bernard Burke of MIT, a pioneer in radio-telescope interferometry who has in recent years become involved in plans for *optical* interferometry, space-based techniques could lead to 10^5-fold increases in resolution over the HST! This is direct imaging, to be sure, but imaging so good that it is in a category by itself.

The idea of the technique is not really new. Albert Michelsen used optical interferometry in the 1920s to measure the diameters of the bright stars Betelgeuse and Antares. Soon thereafter, others used the technique to measure the angular separations of close double stars. The basic principle of optical interferometry is to make use of the information in the *interference pattern* of light waves coming from closely spaced points in the sky. To do this requires the careful manipulation of light (the preserving of *phase* information) from widely spaced telescope mirrors, the longer their separation or "baseline" the better the resolution achievable. In fact, the Very Large Telescope (VLT) to be built at the European Southern Observatory in Chile will be a pioneering optical interferometric array consisting of four eight-meter diameter mirrors. An array of mirrors on the airless Moon could achieve a resolution of 10^{-6} arc second—approximately the angular diameter of Alpha Centauri-A (see Figure 16.3)!

But the most exciting possibility, by far, is to use optical interferometry to do spectroscopic analysis of the atmospheres of extrasolar planets. It might be possible to determine atmospheric composition to find, for example, oxygen: a big clue to the existence of abundant life on a distant

Figure 16.3 Fanciful view of an optical interferometer on the moon. (Courtesy Professor Bernard F. Burke, MIT)

world. Professor Burke is convinced that such measurements could be made in the next century, perhaps even *early* in the new century. How nice that this will happen just when the first robotic interstellar probes might also be ready to travel with a vengeance toward those delectable targets.

Intimations of Other Worlds

With increasing frequency, astronomers in recent years have claimed discovery of planetary systems, formative or protoplanetary systems, and brown dwarfs. In most cases, these have not been confirmed by other observers and have often elicited controversy, but the claims no doubt portend even better *established* results yet to come.

Protoplanetary Systems

Orbiting, airborne, and ground-based infrared telescopes have been used to image dust clouds surrounding certain young stars. According to Beckworth and Sargent, such 100 AU diameter circumstellar disks have been observed around a number of stars, disks believed to be associated with planetary system formation. The Infrared Astronomy Satellite (IRAS) has found more than a dozen infrared-emitting shells around nearby stars. One of the most famous of these, Beta Pictoris, was subsequently imaged in visible light. IRAS also detected a ring of radiating, relatively cold matter around the mature, bright star Vega, a ring of debris that may be akin to the Oort comet cloud in our own Solar System.

The best candidates for planetary systems in the formation process are HL Tau, R Mon, and L 1551 IRS 5 (22). More observations of this phenomenon are to be expected as infrared technology improves. Observation of many of these protoplanetary disks ultimately will allow us to estimate the number of mature planetary systems in the galaxy.

Brown Dwarfs

Using speckle interferometry in the near infrared (at 1.6 and 2.2 microns), D. W. McCarthy, Jr. and his colleagues have reported the discovery of a substellar companion—probably a brown dwarf star—circling the star, Van Biesbroeck 8B or "VB8B" as it has come to be known (23). However, as described by Ronald Schorn, the failure of other observers to confirm the existence of VB8 has led some astronomers to doubt the object's existence (26). The mystery of this "peekaboo" object continues.

The possible discovery of a brown dwarf circling Giclas 29–38, a white dwarf star in Pisces, has recently been reported by Ben Zuckerman and Eric Becklin (24). Using radio spectroscopy and the 1000-foot Arecibo radio telescope in Puerto Rico, a Princeton University group headed by Andrew Fruchter has reported a possible brown dwarf of 23 Jupiter masses circling the pulsar PSR 1957 + 20 (25). Confirmation of any claim to the discovery of a brown dwarf is, however, still lacking.

Epsilon Eridani

Of great significance to starflight enthusiasts is the claim of Bruce Campbell and his colleagues at Dominion Astrophysical Observatory in Canada that they have obtained spectroscopic evidence for a gas giant of 8 to 10 Jupiter masses circling Epsilon Eridani (27). These bodies are much lighter than conventional brown dwarfs. The new techniques enabled the researchers to measure stellar velocities in the radial direction with an accuracy on the order of 10 meters/sec. One of the nearest Sunlike stars, Epsilon Eridani has been a frequent target of searches for alien radio transmissions and is, of course, a tempting destination for our starships.

Postscript

The acceleration in the search for extrasolar planets "intrudes" on the writing of any work with a publication time-scale less than several months. For example, just as this draft was completed in August 1988, a group of astronomers led by David W. Latham at the Smithsonian Astrophysical Observatory in Cambridge, Massachusetts, announced what may be the most conclusive evidence to date of a large planet orbiting a nearby star.

The astronomers detected minute periodic variations in the radial velocity of the star, HD114762, which are apparently caused by the gravitational pull of a large planet with an orbital period of 84 days and a distance from the star about the same as that of Mercury from the Sun. HD114762 is about 90 ly from the Solar System and is a yellowish, medium-size star similar to the Sun, though perhaps twice as old and depleted in metallic elements. Unfortunately, it is not a promising site for life as we know it.

Because of the uncertainty in the inclination of the probable companion's orbit plane to our line of sight, the mass of the object can only be bracketed: probably in the range of 10 to 20 Jupiter masses packed into a similar volume. The astronomers used a spectrograph on the 61-inch reflecting telescope at the Smithsonian's Oak Ridge Observatory in the country town of Harvard, Massachusetts, and used sophisticated mathematical techniques to recover the orbital data from a very noisy signal. But most importantly, their observations were later confirmed by Michel Mayor of the Observatory of Geneva.

Bruce Campbell and his colleagues at the University of Victoria,

British Columbia, using a different spectroscopic observational technique capable of phenomenal velocity resolution (13 meters/sec), have found signs of planetary companions in 8 out of 19 observed nearby stars. In one case, star HR4112 appears to have a companion only about 50% more massive than Jupiter. Our conclusion: starships almost certainly have somewhere to go and go they will!

1 Powers of Ten

Starflight deals not only with numbers that are ponderously large but also very small numbers, as when we consider gossamer thin solar sails. It is important and *simple* to use streamlined scientific notation to deal with such numbers.

For numbers that are *larger* than 1.0:

The superscript tells how many zeros go after the "1."

For example:

10^2 means 100
10^3 means 1000
10^4 means 10,000
10^{11} means 100,000,000,000

etc.

For numbers *smaller* than 1.0:

The superscript with a minus sign indicates how many zeros are in the bottom or denominator of the fraction.

For example:

10^{-1} means 1/10   or 0.1
10^{-2} means 1/100   or 0.01
10^{-3} means 1/1000   or 0.001

10^{-11} means 1/100,000,000,000

etc.

2

Units, Constants, and Physical Data

Units:

Meter	length	(m)
Kilogram	mass	(kg)
Second	time	(sec)
Newton	force	(N)
Watt	power	(W)
Joule	energy	(J)
Coulomb	charge	(C)
Electron Volt	energy	(eV)
Kilo Electron Volt	energy	(KeV)
Henry	inductance	(H)
Farad	capacitance	(F)
Arc second	angle	—
Parsec	length	—
Light year	length	(ly)
Astronomical unit	length	(AU)
Hertz	frequency	(Hz)
Kelvin degree	temperature	(K)
Angstrom	length	(Å)

Physical Constants[1]:

Boltzmann constant	k	1.380658×10^{-23}	J/K
Electron charge	e	$1.60217733 \times 10^{-19}$	C
Electron charge/mass ratio	e/m_e	$1.75881962 \times 10^{11}$	C/kg
Electron rest mass	m_e	$0.91093897 \times 10^{-30}$	kg
Electron rest mass/proton rest mass	m_e/m_p	$5.44617013 \times 10^{-4}$	—
Gravitational acceleration (Earth surface, typical value)	g	9.81	m/sec^2
Gravitational constant	G	6.67206×10^{-11}	Nm^2/kg^2
Light speed in vacuo	c	299792458	m/sec
Neutron rest mass	m_n	$1.6749286 \times 10^{-27}$	kg
Permeability constant	μ_0	$12.5663706 \times 10^{-7}$	H/m

Permittivity constant	e_0	$8.854187817 \times 10^{-12}$	F/m
Planck's constant	h	$6.6260755 \times 10^{-34}$	J/Hz
Proton mass/electron mass	m_p/m_e	1836.152701	—
Proton rest mass	m_p	$1.6726231 \times 10^{-27}$	kg
Stefan-Boltzmann constant	σ	5.67051×10^{-8}	$W/(m^2 K^4)$

[1]Emiliani, Cesare, *The Scientific Companion: Exploring the Physical World with Facts, Figures, and Formulas*, New York: John Wiley & Sons, Inc., 1988.

Numeric constants:

Pi	π	3.1415926536
Natural base	e	2.7182818285

Physical Data:

Mass of the Sun	1.99×10^{30}kg
Radius of Sun	6.96×10^{8}m
Mass of the Earth	5.98×10^{24}kg
Radius of Earth	6.378×10^{6}m
Solar flux at 1 AU	≈ 1400 watts/m^2

Nearby Star Systems

Star System Number	Star Designation	Component	Distance from Sol (light years)	Right Ascension α (hours).(minutes)	Declination δ (degrees).(min.)
0	Sol	—	—	—	—
1	Proxima Centauri	—	4.3	14.26	− 62.28
2	α Centauri; 128620	A	4.4	14.36	− 60.38
	α Centauri; 128620	B	4.4	14.36	− 60.38
3	Barnard's Star; (+ 4° 3561)	—	5.9	17.55	4.33
4	Wolf 359	—	7.6	10.54	7.19
5	Lalande 21185; BD + 36° 2147	—	8.1	11.01	36.18
6	Sirius; 48915	A	8.7	6.43	− 16.39
	Sirius; 48915	B	8.7	6.43	− 16.39
7	Luyten 726-8	A	8.9	1.36	− 18.13
	UV Ceti	B	8.9	1.36	− 18.13
8	Ross 154; AC-242833-183	—	9.5	18.47	− 23.53
9	Ross 248	—	10.3	23.39	43.55
10	ε Eridani; 22049	—	10.7	3.31	− 9.38
11	Luyten 789-6	—	10.8	22.36	− 15.36
12	Ross 128		10.8	11.45	1.06
13	61 Cygni; 201091	A	11.2	21.05	38.30
	61 Cygni; 201092	B	11.2	21.05	38.30
14	ε Indi; 209100	—	11.2	22.00	− 57.00
15	Procyon; 61421; α Canis Minoris	A	11.4	7.37	5.21
	Procyon; 61421; α Canis Minoris	B	11.4	7.37	5.21
16	+ 59° 1915; 173739	A	11.5	18.42	59.33
	+ 59° 1915; 173740	B	11.5	18.42	59.33
17	Groombridge 34; BD + 43°44; 1326	A	11.6	0.15	43.44
	Groombridge 34; BD + 43°44; 1326	B	11.6	0.15	43.44
18	Lacaille 9352; CD − 36° 15693	—	11.7	23.03	− 36.08
19	τ Ceti	—	11.9	1.42	− 16.12
20	Luyten BD + 5° 1668	—	12.2	7.25	5.23
21	L725-32; LET 118	—	12.5	1.07	− 17.32
22	Lacaille 8760; CD − 39° 14192; 202560	—	12.5	21.14	− 39.04
23	Kapteyn's Star; − 45° 1841	—	12.7	5.10	− 45.00

Equatorial System Coordinates				Luminosity	Mass	Radius	Proper Motion	Radial Velocity	Angle of Proper Motion From North
x (ly)	y (ly)	z (ly)	Spectral Type	$L/L_\odot$	$M/M_\odot$	$R/R_\odot$	(sec/yr)	(km/sec)	(degrees)
0.0	0.0	0.0	G2	1.0	1.0	1.0	N/A	N/A	—
− 1.6	− 1.2	− 3.8	M5e	0.00006	0.1	—	3.85	− 16	282
− 1.7	− 1.4	− 3.8	G2	1.3	1.10	1.23	3.68	− 22	281
− 1.7	− 1.4	− 3.8	K6	0.36	0.89	0.87	3.68	− 22	281
− 0.1	− 5.9	0.5	M5	0.00044	0.15	0.12	10.31	− 108	356
− 7.2	2.1	1.0	M8e	0.00002	0.20	0.04	4.71	+ 13	235
− 6.3	1.7	4.8	M2	0.0052	0.35	0.35	4.78	− 84	187
− 1.6	8.2	− 2.5	A1	23.0	2.31	1.8	1.33	− 8	204
− 1.6	8.2	− 2.5	DA	0.0028	0.98	0.022	1.33	− 8	204
7.7	3.4	− 2.8	M6e	0.00006	0.12	0.05	3.36	+ 30	80
7.7	3.4	− 2.8	M6e	0.00004	0.10	0.04	3.36	+ 32	80
1.8	− 8.5	− 3.8	M5e	0.0004	0.31	0.12	0.72	− 4	103
7.4	− 0.7	7.1	M6e	0.00011	0.25	0.07	1.58	− 81	176
6.4	8.4	− 1.8	K2	0.30	0.8	0.90	0.98	+ 16	271
9.7	− 3.7	− 2.9	M6	0.00012	0.25	0.08	3.26	− 60	46
− 10.8	0.7	0.2	M5	0.00033	0.31	0.10	1.37	− 13	153
6.3	− 6.1	7.0	K5	0.063	0.59	0.70	5.22	− 65	52
6.3	− 6.1	7.0	K7	0.040	0.50	0.80	5.22	− 63	52
5.3	− 3.0	− 9.4	K5	0.13	0.71	1.0	4.69	− 40	123
− 4.7	10.3	1.1	F5	7.6	1.77	1.7	1.25	− 3	214
− 4.7	10.3	1.1	DF	0.0005	0.63	0.01	1.25	—	214
1.1	− 5.7	9.9	M4	0.0028	0.4	0.28	2.28	0	324
1.1	− 5.7	9.9	M5	0.0013	0.4	0.20	2.28	+ 10	324
8.4	0.5	8.0	M2 (spectro-scopic double)	0.0058	0.38	0.38	2.89	+ 13	89
8.4	0.5	8.0	M4	0.0004	—	0.11	2.89	+ 20	82
9.2	− 2.3	− 6.9	M2	0.012	0.47	0.57	6.90	+ 10	79
10.3	4.9	− 3.3	G8	0.44	0.82	1.67	1.92	− 16	292
− 4.4	11.3	1.1	M4	0.0014	0.38	0.16	3.73	+ 26	171
11.4	3.4	− 3.8	M5e	—	—	—	—	—	—
7.3	− 6.4	− 7.9	M1	0.025	0.54	0.82	3.46	+ 21	251
1.9	8.8	− 9.0	M0	0.004	0.44	0.24	8.89	+ 245	131

Nearby Star Systems

Star System Number	Star Designation	Component	Distance from Sol (light years)	Right Ascension α (hours).(minutes)	Declination δ (degrees).(min.)
24	Krüger 60; 239960	A	12.8	22.26	57.27
	DO Cephei; 239960	B	12.8	22.26	57.27
25	Ross 614	A	13.1	6.27	− 2.46
	Ross 614	B	13.1	6.27	− 2.46
26	BD − 12° 4523	—	13.1	16.28	− 12.32
27	van Maanen's Star; Wolf 28	—	13.9	0.46	5.09
28	Wolf 424	A	14.2	12.31	9.18
	Wolf 424	B	14.2	12.31	9.18
29	G158-27		14.4	0.04	− 7.48
30	CD − 37° 15492	—	14.5	0.02	− 37.36
31	Groombridge 1618; BD + 50° 1725	—	15.0	10.08	49.42
32	CD-46° 11540	—	15.1	17.25	− 46.51
33	CD-49° 13515	—	15.2	21.30	− 49.13
34	CD-44° 11909	—	15.3	17.33	− 44.17
35	Luyten 1159-16	—	15.4	1.57	12.51
36	Lalande 25372; BD + 15° 2620; 119850	—	I5.7	13.43	15.10
37	BD + 68° 946; AOe 17415-6	—	15.8	17.37	68.23
38	Luyten 145-141; CC658	—	15.8	11.43	− 64.33
39	Ross 780; BD-15° 6290	—	15.8	22.51	− 14.31
40	40 Eridani; Omicron Eridani; 26965	A	15.9	4.13	− 7.44
	40 Eridani; − 7° 781; 26976	B	15.9	4.13	− 7.44
	40 Eridani; − 7° 781; 26976	C	15.9	4.13	− 7.44
41	BD + 20° 2465	—	16.1	10.17	20.07
42	Altair; 187642	—	16.6	19.48	8.44
43	70 Ophiuchi; + 2° 3482	A	16.7	18.03	2.31
	70 Ophiuchi; 165341	B	16.7	18.03	2.31
44	AC + 79° 3888	—	16.8	11.45	78.58
45	BD + 43° 4305	—	16.9	22.45	44.05
46	Stein 2051; AC + 58 25001	A	17.0	4.26	58.53
	Stein 2051; AC + 58 25002	B	17.0	4.26	58.53
47	+ 44° 2051	A	17.5	11.03	43.47
	WX Ursa Majoris	B	17.5	11.03	43.47
48	− 26° 12026; 155886	A	17.7	17.12	− 26.32
	36 Ophiuchi; 155885	B	17.7	17.12	− 26.32
	− 26° 12036; 156026	C	17.7	17.12	− 26.32

| Equatorial System Coordinates | | | | Luminosity | Mass | Radius | Proper Motion | Radial Velocity | Angle of Proper Motion From North |
x (ly)	y (ly)	z (ly)	Spectral Type	$L/L_\odot$	$M/M_\odot$	$R/R_\odot$	(sec/yr)	(km/sec)	(degrees)
6.3	− 2.7	10.8	M4	0.0017	0.27	0.51	0.86	− 26	245
6.3	− 2.7	10.8	M6	0.00044	0.16	—	0.90	− 26	245
− 1.5	13.0	− 0.6	M5e	0.0004	0.14	0.14	0.99	+ 24	134
− 1.5	13.0	− 0.6	—	0.00002	0.08	—	0.99	+ 24	134
− 5.0	− 11.8	− 2.8	M5 (spectro-scopic double)	0.0013	0.38	0.22	1.18	− 13	182
13.6	2.8	1.2	DG	0.00017	—	—	2.95	+ 54	155
− 13.9	− 1.9	2.3	M6e	0.00014	—	0.09	1.75	− 5	277
− 13.9	− 1.9	2.3	M6e	0.00014	—	0.09	1.75	− 5	277
14.3	0.2	− 2.0	M	0.00005	—	—	2.06	—	204
11.5	0.1	− 8.8	M3	0.00058	0.39	0.4	6.08	+ 23	113
− 8.6	4.6	11.4	K7	0.04	0.56	0.5	1.45	− 26	249
− 1.6	− 10.2	− 11.0	M4	0.003	0.44	0.25	1.13	—	147
7.9	− 6.0	− 11.5	M3	0.0058	0.37	0.34	0.81	+ 8	185
− 1.3	− 10.9	− 10.7	M5	0.00063	0.34	0.15	1.16	—	217
13.1	7.3	3.4	M8	0.00023	—	—	2.08	—	149
− 13.6	− 6.6	4.1	M2	0.0076	—	0.40	2.30	+ 15	129
− 0.6	− 5.8	14.7	M3	0.0044	0.35	0.39	1.33	− 22	194
− 6.8	0.5	− 14.3	DA	0.0008	—	—	2.68	—	97
14.6	− 4.5	− 4.0	M5	0.0016	0.39	0.23	1.16	+ 9	125
7.1	14.1	− 2.1	K0	0.33	0.11	0.7	4.08	− 43	213
7.1	14.1	− 2.1	DA	0.0027	0.43	0.018	4.11	− 21	213
7.1	14.1	− 2.1	M4e	0.00063	0.21	0.43	4.11	− 45	213
− 13.6	6.6	5.5	M4	0.0036	0.44	0.28	0.49	+ 11	264
7.4	− 14.6	2.5	A7	10.0	590	1.2	0.66	− 26	54
0.2	− 16.7	0.7	K1 (spectro-scopic double)	0.44	0.89	1.3	1.12	− 7	167
0.2	− 16.7	0.7	K6	0.083	0.68	0.84	1.12	− 10	167
− 3.2	0.2	16.5	M4	0.0009	0.35	0.15	0.89	− 119	57
11.5	− 3.9	11.8	M5e	0.0021	0.26	0.24	0.83	− 2	237
3.5	8.1	14.6	M5	0.0008	—	—	2.37	—	146
3.5	8.1	14.6	DC	0.0003	—	—	2.37	—	146
− 12.2	3.1	12.1	M2	—	—	—	4.54	+ 65	—
− 12.2	3.1	12.1	M8	—	—	—	4.54	+ 65	—
− 3.3	− 15.5	− 7.9	K2	0.26	0.77	0.90	1.24	− 1	—
− 3.3	− 15.5	− 7.9	K1	0.26	0.76	0.82	1.23	0	—
− 3.3	− 15.5	− 7.9	K6	0.09	0.63	0.90	1.22	− 1	—

Nearby Star Systems

Star System Number	Star Designation	Component	Distance from Sol (light years)	Right Ascension α (hours).(minutes)	Declination δ (degrees).(min.)
49	− 36° 13940; HR 7703; 191408	A	18.4	20.08	− 36.14
	− 36° 13940; 191408	B	18.4	20.08	− 36.14
50	σ Draconis; 185144	—	18.5	19.32	69.35
51	Ross 882; YZ Canis Minoris	—	18.5	7.40	3.48
52	δ Pavonis; 190248	—	18.6	20.04	− 66.19
53	1° 4774	—	18.6	23.47	2.08
54	Luyten 347-14	—	18.6	19.17	− 45.37
55	− 21° 1377; 42581	—	18.7	6.08	− 21.51
56	Luyten 97-12	—	18.9	7.53	− 67.38
57	Luyten 674-15	—	19.1	8.10	− 21.24
58	η Cassiopeia; 4614	A	19.2	0.46	57.33
	η Cassiopeia; 4614	B	19.2	0.46	57.33
59	Luyten 205-128; UC 48	—	19.2	17.42	− 57.17
60	− 3° 1123; HD 36395	—	19.2	5.29	− 3.41
61	− 40° 9712	—	19.3	15.29	− 41.06
62	Ross 986; AC + 38 23616	—	19.3	7.07	38.38
63	Ross 47; AC + 12 1800-213	—	19.4	5.39	12.29
64	Wolf 294; AC + 33 25644	—	19.4	6.52	33.20
65	LP 658-2	—	19.6	5.53	− 4.08
66	+ 53° 1320; 79211	A	19.6	9.11	52.54
	+ 53° 1321; 79210	B	19.6	9.11	52.54
67	+ 4° 4048; 180617	A	19.6	19.14	5.06
	+ 4° 4048; VB10	B	19.6	19.14	5.06
68	− 45° 13677; 191849	—	19.9	20.10	− 45.19
69	82 Eridani; 20794	—	20.3	3.17	− 43.16
70	Wolf 630; − 8° 4352	A	20.3	16.53	− 8.15
	Wolf 630; − 8° 4352	B	20.3	16.53	− 8.15
	VB 8	C	20.3	16.53	− 8.15
	Wolf 629	D	20.3	16.53	- 8.14
71	− 11° 3759	—	20.4	14.32	− 12.19
72	β Hydri; 2151	—	20.5	0.23	− 77.32
73	+ 45 2505; 155876	A	21.0	17.11	45.45
	+ 45 Fu46; 155876	B	21.0	17.11	45.45
74	+ 19° 5116	A	21.0	23.20	19.40
	+ 19° 5116	B	21.0	23.20	19.40

Equatorial System Coordinates				Luminosity	Mass	Radius	Proper Motion	Radial Velocity	Angle of Proper Motion From North
x (ly)	y (ly)	z (ly)	Spectral Type	$L/L_\odot$	$M/M_\odot$	$R/R_\odot$	(sec/yr)	(km/sec)	(degrees)
7.9	− 12.6	− 10.9	K3	0.20	0.76	0.80	1.65	—	—
7.9	− 12.6	− 10.9	M5	0.0008	0.35	0.14	1.65	− 130	—
2.5	− 5.9	17.3	K0	0.4	0.82	0.28	1.83	+ 27	—
− 7.8	16.7	1.2	M4	—	—	—	0.61	+ 18	—
3.8	− 6.4	− 17.0	G6	1.0	0.98	1.07	1.65	− 22	—
18.6	− 1.1	0.7	M2	0.0001	—	—	1.59	− 65	—
4.3	− 12.3	− 13.3	M7	0.0001	0.26	0.08	2.94	—	—
− 0.6	17.3	− 7.0	M1	0.016	0.46	0.59	0.74	+ 4	—
− 3.4	6.3	− 17.5	D	0.0003	—	—	2.05	—	—
− 9.6	15.0	− 7.0	M	—	—	—	0.73	—	—
10.1	2.1	16.2	G0	1.0	0.85	0.84	1.11	+ 9	—
10.1	2.1	16.2	M0	0.03	0.52	0.07	1.11	+ 13	—
− 0.8	− 10.3	− 16.2	M	0.0002	0.14	—	1.7	—	—
2.6	19.0	− 1.2	M1	0.02	0.51	0.69	2.23	+ 11	—
− 8.9	− 11.5	− 12.7	M4	0.003	0.44	0.29	1.55	—	—
− 4.3	14.4	12.0	M5	—	—	—	1.08	+ 39	—
1.7	18.9	4.2	M6	0.0008	0.35	0.17	2.37	+ 103	—
− 3.6	15.8	10.7	M4	0.008	0.49	0.46	2.37	—	—
0.6	19.5	− 1.4	DK	—	—	—	2.37	—	—
− 8.8	7.9	15.6	M0	—	—	—	1.68	+ 11	—
− 8.8	7.9	15.6	M0	—	—	—	1.70	+ 10	—
6.2	− 18.5	1.7	M4	—	0.39	0.43	1.46	+ 33	—
6.2	− 18.5	1.7	M5	0.007	—	0.008	1.49	+ 33	—
7.5	− 11.8	− 14.1	M0	0.00002	—	—	0.78	− 30	—
9.6	11.2	− 13.9	G5	—	—	—	3.12	+ 87	—
− 5.8	− 19.2	− 2.9	M4	—	0.38	—	1.18	+ 19	—
− 5.8	− 19.2	− 2.9	M5	—	0.38	—	1.18	+ 19	—
− 5.8	− 19.2	− 2.9	—	—	—	—	1.18	+ 19	—
− 5.8	− 19.2	− 2.9	M4 (spectro-scopic double)	—	—	—	1.19	+ 22	—
− 15.7	− 12.3	− 4.4	M4	—	—	—	0.69	—	—
4.4	0.4	− 20.0	G1	—	—	1.66	2.25	+ 23	—
− 3.1	− 14.3	15.0	M3	—	0.31	—	1.59	− 21	—
− 3.1	− 14.3	15.0	—	—	0.25	—	1.59	− 21	—
19.5	− 3.4	7.1	M4	—	—	—	0.55	− 1	—
19.5	− 3.4	7.1	M6	—	—	—	0.55	− 4	—

4 Guide to Starflight Literature

If one were stranded on a rocky asteroid and granted the wish to have but one source of information about starflight, it would be impossible not to pick the "red cover" issues of the *Journal of the British Interplanetary Society*. Since they first appeared in 1970 (as the "New Frontiers" series), these publications have served virtually as the world focus of interstellar studies. For nearly two decades these special issues of the Society's journal have sparked the imaginations of those interested in starflight, extraterrestrial life, and the search for extraterrestrial intelligence. Before the appearance of the *Journal*'s first "interstellar studies" issue (April 1974), the Society's other monthly publication, *Spaceflight*, was the repository of many articles on starflight and SETI. *Spaceflight* no longer has those themes as a major focus, but it continues to be one of the world's best general references on space exploration and development.

The *Journal of the British Interplanetary Society* and *Spaceflight* are not likely to be found in the magazine section of your local supermarket, so you will have to look for them, typically in university libraries. Dyed-in-the-wool starflight pioneers who may wish to subscribe should write to:

Editorial Office
The British Interplanetary Society
27/29 South Lambeth Road
London, SW8 1SZ, England

An especially useful feature that *JBIS* has published is the compendious bibliography of starflight and SETI and its subsequent updates:

"Interstellar Travel and Communication: A Bibliography." Eugene F. Mallove, Robert L. Forward, Zbigniew Paprotny, and Jurgen Lehmann. *JBIS* 33 (June 1980): 47 pages (entire issue), 2699 references.

"Interstellar Travel and Communication Bibliography: 1982 Update." Zbigniew Paprotny and Jurgen Lehmann. *JBIS* 36 (July 1983): 311–329, 750 references.

"Interstellar Travel and Communication Bibliography: 1984 Update." Zbigniew Paprotny, Jurgen Lehmann, and John Prytz. *JBIS* 37 (November 1984): 502–512, 644 references.

"Interstellar Travel and Communication Bibliography: 1985 Update." Zbigniew Paprotny, Jurgen Lehmann, and John Prytz. *JBIS* 39 (March 1986): 127–136, 572 references.

Another publication that has explored interstellar themes for many years is the planetary science journal, *Icarus. Icarus* is often available in college geophysics or aerospace libraries. Write to:

Dr. Joseph A. Burns, Editor
Icarus
Space Sciences Building
Cornell University
Ithaca, NY 14853

The American Institute for Aeronautics and Astronautics, now based in Washington (370 L'Enfant Promenade SW, Washington, D.C., 20024), has also generously opened its series of technical papers in astronautics to discussions of interstellar flight. Likewise have the American Astronautical Society (AAS) and the International Astronautical Federation (IAF).

Four indispensable organizations should not be forgotten:

• The Planetary Society, the international space exploration and SETI advocacy group based in the U.S. The organization's profusely illustrated monthly newsletter and the group's research activities are worth staying in touch with. Write: The Planetary Society, 65 N. Catalina Avenue, Pasadena, CA 91106.

• The Space Studies Institute, a research organization founded by Gerard O'Neill to promote the colonization of space and the utilization of extraterrestrial resources. Write: The Space Studies Institute, 285 Rosedale Road, P.O. Box 82, Princeton, New Jersey 08540.

• The World Space Foundation, a nonprofit research organization that has led in the development of solar sails. Write: World Space Foundation, P.O. Box Y, South Pasadena, CA 91030–1000.

• The International Space University, a multinational project (founded at MIT in 1987) designed to create a worldwide center for training tomorrow's space professionals. Write: International Space University, 636 Beacon Street, Suites 201–202, Boston, MA 02215.

Few books have treated interstellar flight in more than very general terms, and virtually none until this work has presented starflight in its broadest scope. Still, there are landmark books sufficiently memorable to warrant mention. Some of these are as much or more devoted to SETI but contain historic references to starflight. This is our admittedly eclectic selection:

Realities of Space Travel: Selected Papers of the British Interplanetary Society. Edited by L.J. Carter. New York: McGraw-Hill Book Company, 1957.

Interstellar Communication: The Search for Extraterrestrial Life. Edited by A.G.W. Cameron. New York: W.A. Benjamin, Inc., 1963.

Flight to the Stars: An Inquiry into the Feasibility of Interstellar Flight. James Strong. New York: Hart Publishing Company, Inc., 1965.

Intelligent Life in the Universe. I.S. Shklovskii and Carl Sagan. San Francisco: Holden-Day, Inc., 1966.

Intelligence in the Universe. Roger A. MacGowan and Frederick Ordway, III. Englewood Cliffs, NJ: Prentice-Hall, Inc., 1966.

How We Will Reach the Stars. John W. Macvey. New York: Collier Books, 1969.

Extraterrestrial Civilizations: Problems of Interstellar Communication. Edited by S.A. Kaplan. Moscow: 1969. Translated from Russian by The Israel Program for Scientific Translations, 1971.

Communication with Extraterrestrial Intelligence (CETI). Edited by Carl Sagan. Cambridge, MA: MIT Press, 1973.

The Galactic Club: Intelligent Life in Outer Space. Ronald N. Bracewell. San Francisco: W.H. Freeman and Company, 1974.

Interstellar Communication: Scientific Perspectives. Edited by Cyril Ponnamperuma and A.G.W. Cameron. Boston: Houghton Mifflin Company, 1974.

Worlds Beyond. Ian Ridpath. London: Wildwood House, 1975.

Migration to the Stars. Edward S. Gilfillan, Jr. Washington, DC: Robert B. Luce Co., Inc., 1975.

The Road to the Stars. Iain Nicolson. New York: Morrow, 1978.

The Search for Life in the Universe. Donald Goldsmith and Tobias Owen. Menlo Park, CA: The Benjamin/Cummings Publishing Company, Inc., 1980.

The Quest for Extraterrestrial Life: A Book of Readings. Donald Goldsmith. Mill Valley, CA: University Science Books, 1980.

Bound for the Stars. Saul J. Adelman and Benjamin Adelman. Englewood Cliffs, NJ: Prentice-Hall, Inc., 1981.

The Coattails of God. R.M. Powers. New York: Warner, 1981.

Contact with the Stars. Reinhard Breuer. Oxford and San Francisco: Freeman, 1982.

Interstellar Migration and the Human Experience. Edited by Ben R. Finney and Eric M. Jones. Berkeley: University of California Press, 1985.

The Search for Extraterrestrial Intelligence: Listening for Life in the Cosmos. Thomas R. McDonough. New York: John Wiley and Sons, Inc., 1987.

Mirror Matter: Pioneering Antimatter Physics. Robert L. Forward and Joel Davis. New York: John Wiley & Sons, Inc., 1988.

Starsailing: Solar Sails and Interstellar Travel. Louis Friedman. New York: John Wiley & Sons, Inc., 1988.

The Twin "Paradox"

It is an ancient tale in astronautics, so old that its origins have nearly been forgotten: the paradox of twins. Each version of the story goes something like this: Alice and Jill, identical twin astronauts in the Earthian Space Program, kiss each other good-bye on their 30th birthday. Jill is about to depart the TerraTwo space colony (orbiting the Sun) in her new compact model Bussard interstellar ramjet, the Millennium Squid. Jill blasts off and accelerates to near-light speed in less than a year, beginning a long cruise phase during which her velocity relative to the Solar System remains constant. After exploring countless light years of boring interstellar space, homesick Jill decides to return to TerraTwo. So she decelerates her ramjet to zero velocity (relative to the Solar System), turns her ship around, accelerates back up to near-light speed, and cruises back along all those boring light years of space. Nearing the Solar System, Jill decelerates the ramjet, and approaches the TerraTwo space colony.

To her dismay, but not astonishment, 40-year-old Jill finds the colony abandoned and in decay. Thousands of years of erosion by interplanetary debris has turned the once vibrant space city into a useless hulk. Jill surmises that thousands or perhaps millions of years have gone by for TerraTwo, yet she has only a few more facial wrinkles to show for her decade-long space cruise. Her sister Alice died centuries ago, as Jill easily determines when she consults the still working computerized colony log and compares Alice's recorded time of death with TerraTwo's lone surviving atomic calendar clock. (Obviously, the two astronauts do not have to be twins for there to be a "paradox." For example, husband and wife would also do. But twins accentuate the paradoxical quality. And they are in a long . . . Tradition!)

How can this be? Even though we know from Einstein's theory of Special Relativity that "common sense time"—the universal "nowness" of time—was not what we thought it to be, we are still puzzled by this nonintuitive situation. Of course, we had learned from Einstein that two people moving with respect to each other at high-constant velocity would each observe that the other's clocks (and all other processes, including life itself) had apparently slowed down. This was a perfectly "acceptable" paradox because neither observer would have to confront the other face-to-face to reconcile the conflict. But with the twin story, we have what appears to be a similarly symmetrical situation, and yet we see the "paradox" that one twin ages little while the other ages a lot, in fact, has been dead a long time. True, one twin accelerates and decelerates during brief periods, but for much of her time she is cruising at a steady velocity relative to the Solar System. Why should these brief periods of acceleration and deceleration make any difference in an otherwise symmetrical situation? Of this "logical impasse" Bernard Oliver has written, "The student concludes that either he is out of his mind (which is distressing) or Einstein was (which is irreverent, but less distressing). In this way, much skepticism of relativity develops"[1].

The answer is that acceleration *does* count! The twins are by no means symmetrical or equivalent observers. Jill experiences acceleration from the propulsive maneuvers, while her sister Alice in TerraTwo does not. By accelerating, Alice has, in effect, switched between inertial (*nonaccelerated* frames of reference) *four* times. It is really possible to travel into the future, though at great expense and trouble, and only if you are willing to say good-bye to loved ones left back home. No part of Special Relativity has caused so much controversy, and the literature is replete with debates among physicists as well as the less well informed. (Consult the many references on the twin paradox that are listed in the *Bibliography on Interstellar Travel and Communication*, referenced in Appendix 4.) Is there a way to get some intuitive grasp about why the twin "paradox" is no paradox at all? Fortunately, yes. A number of commentators have provided nice graphical or simple calculational means of understanding how the twin story works [1,2].

First, what is the "dilation" of Jill's time that needs explaining? If her journey is symmetrical in acceleration and deceleration, outbound and returning, the TerraTwo colony will observe her round-trip time to be T with each leg taking T/2. (For simplicity, also assume in what follows that the periods of acceleration and deceleration take negligible time compared with the remainder of the trip, a perfectly good assumption that detracts little from the basic conclusion, as could be shown by direct calculation.) The dilated, slowed, or *compressed trip time experienced by Jill will be T'*. According to Relativity:

$$T' = \frac{T}{\gamma} \text{ or } T = \gamma T'$$

where, the dilation factor, gamma (see Chapter 3), is:

$$\gamma = \frac{1}{\sqrt{1 - \dfrac{V^2}{c^2}}}$$

in which V is the relative velocity of the two observers during cruise and c is the velocity of light. If, for example, V = .999 c, then T = 22.4 T'; for V = 0.999999, then T = 707 T', and so on.

Take a much less extreme example of time compression—flight at 0.6 c —to examine one of the most insightful graphical depictions of the twin paradox we have found, shown in Figure 1, which is adapted from Bernard Oliver's paper (1). V = 0.6 c is such that the time in the stationary frame (call this "Earth") is 1.25 times the starship frame (instantaneous accelerations are assumed at turnaround points), following simply from the expression above. Figure 1 is a space-time diagram for the movement of the 0.6 c starship through time and space (tilted path)

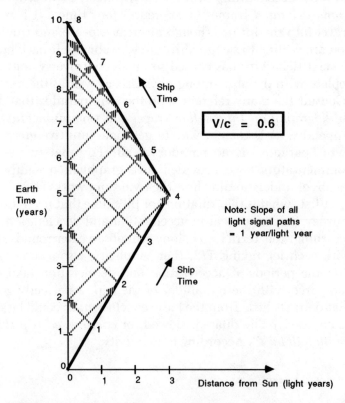

Figure 1. Space—time diagram of the twin effect (0.6c). (Courtesy Dr. Bernard M. Oliver and IEEE)

and the path of the Earth observer through time (vertical path). The starship makes a 3-ly trip outbound at 0.6 c, turns around, and comes back to Earth at 0.6 c, a total time clocked by the Earth observer as $T = 3/0.6 + 3/0.6 = 10$ years. So by formula the starship clock should read 8 years upon its return to Earth.

Before the starship departs, it agrees to send out one radio pulse per year (of its shipboard time) back to Earth. Likewise, the Earth agrees to send one pulse each year to the starship. The paths of the pulses are depicted by the two different dotted line styles shown in the diagram. Note well: Pulse paths, whether from the Earth or starship, are always 45° to each axis because they travel with the speed of light (1 ly per year).

Note that on the outbound leg, the starship receives only 2 pulses from the Earth, while on its return flight it receives 8 pulses from Earth—a total of 10 pulses. Observe, however, that Earth receives only 4 pulses from the starship on its outbound flight (during an Earth period of 8 years) and receives 4 pulses from the starship on its return flight (during an Earth period of 2 years). The last starship and Earth pulses occur just as the starship lands on Earth. Each observer *is* sending out one pulse per year of its time, yet pulses are being received by the starship and Earth in the ratio, 10 to 8 or 1.25! Clearly there is a slowing of time for the starship compared with Earth time. The different pulse rates received by the starship and Earth correspond to observed Doppler shifts in their frequencies, as indicated in the diagram.

If the starship were traveling much closer to the speed of light, its space-time paths—outbound and inbound—would be much closer to 45° to each axis, and difference in the frequency and total number of pulse arrivals between frames would be even more dramatic. To be sure, some people will refuse to be convinced and will continue to doubt the reality of what should really be called the "Twin Effect."

Energy, Efficiency, and Starflight Propulsion

It is an inescapable fact that to propel a mass to a certain velocity requires the input of energy, which then becomes part of the kinetic energy of the object. If the body is massive and the velocity high, the cost of the energy input might well reach a significant portion of the annual energy expenditure of civilization for all its pursuits. Many skeptics have criticized starflight partially on the grounds that we could never afford the energy necessary to carry it out. For example, Bernard Oliver wrote, "Even if intelligent life is common in the universe, it should come as no surprise that we have not been visited. Perhaps their appropriations committees are just as reluctant as ours to finance projects that would take thousands of years or that require thousands of years of their planet's energy needs. Perhaps they are intelligent enough to realize how much cheaper, easier, and safer interstellar communication is, and how nearly equally rewarding"(3). (Dr. Oliver comes to this conclusion by denying the feasibility or desirability of worldships, assuming the need for high-speed relativistic flights, insisting on two-way flight, and ignoring the possibility and utility of low-mass scientific payloads.) So it is, indeed, worthwhile to put energy usage in perspective by relating the energy "cost" of interstellar flight to energy availability. The kinetic energy (K) of a body at nonrelativistic speeds is the familiar expression:

$$K = \frac{1}{2}mV^2$$

An exact expression for kinetic energy, according to Special Relativity, is the difference between the particle's total energy, E, and the rest energy, E_o:

$$K = E - E_o$$

or

$$K = \frac{m_0 c^2}{\sqrt{1 - \dfrac{V^2}{c^2}}} - m_0 c^2$$

The latter expression reduces, of course, to the nonrelativistic form when $V^2/c^2 << 1.0$.

The kinetic energy of a payload may be viewed as the *minimum* required energy expenditure to boost it to velocity, V. This is because the process necessary to boost the payload (e.g. rocket propulsion) is by no means perfectly efficient in translating released energy to payload kinetic energy. Since different propulsion systems have various efficiencies, first simplify the discussion by postulating perfect energy conversion by the system.

Think of the kinetic energy on a per unit mass basis to get some idea of how much energy input must go into boosting a payload to a particular fraction of the velocity of light. Table 1 presents values for this relationship.

Table 1.   Kinetic Energy vs. Velocity

V/c	K/m_0 (Kinetic energy per kilogram of rest mass; Joules/kg)
0.001	4.5×10^{10}
0.005	1.1×10^{12}
0.01	4.5×10^{12}
0.02	1.8×10^{13}
0.05	1.1×10^{14}
0.10	4.5×10^{14}
0.15	1.0×10^{15}
0.20	1.8×10^{15}
0.25	3.0×10^{15}
0.30	4.3×10^{15}
0.35	6.1×10^{15}
0.40	8.2×10^{15}
0.50	1.4×10^{16}
0.60	2.3×10^{16}
0.70	3.6×10^{16}
0.80	6.0×10^{16}
0.90	1.2×10^{17}
0.95	2.0×10^{17}
0.99	5.5×10^{17}

These are very large numbers, but how do they compare with known quantities? Table 2 gives some interesting comparisons.

Table 2. Energy Equivalents

	Energy (Joules)
Solar electromagnetic energy flux penetrating sphere of radius 1 AU in *one year*	1.2×10^{34}
Solar energy flux on Earth's cross section in *one year*	$5.3. \times 10^{24}$
Kinetic energy of an asteroid $\approx$ 2 km diameter; 10^{10} tons; 30km/sec	4.5×10^{21}
Solar energy flux on Earth's cross section in *one minute*	1.0×10^{19}
U.S. annual energy consumption (1970)	7×10^{18}
U.S. annual electrical power (1982)	8.5×10^{18}
World annual electrical power (1982)	3.1×10^{19}
Energy release by one Saturn V rocket	10^{13}

From this table, it is safe to estimate the annual world civilization *total* energy expenditure in the present era to be on the order of 10^{19} joules. 10^{19} joules is defined as an energy unit of one "GAEE," Global Annual Energy Expenditure (pronounced "gee," but looks suspiciously like Gaia). How much mass at what speed can be launched into interstellar space with one GAEE of energy? Consider three different payload categories: (1) 1 kilogram (Starwisp class vehicle, Chapter 5); (2) 10^5 kilograms (Daedalus class payload, Chapter 4); (3) 10^7 kilograms (Sark-1 class interstellar ark, Chapter 6).

One GAEE could dispatch 10,000 one-kilogram probes at 0.15 c, a single Daedalus-class 10^5 kilogram payload at 0.05 c, or a 1000-person interstellar ark at 0.005 c. Note that one GAEE is about the solar energy flux falling on Earth during one minute! As our civilization develops into a truly spacefaring culture, there will be the deployment of increasing areas of solar energy collectors in space, which will eventually equal significant fractions of the Earth's cross-sectional area. One GAEE of convertible solar power could well become available for starflight within the next two centuries. It might take more than one minute to gather it, if our collectors do not become as big as the Earth. But it will be worth waiting a few days, weeks, or months for the prize of interstellar voyages powered ultimately by "free" sunlight. And one GAEE is *just the beginning* of what a truly interplanetary civilization might eventually command. Think of what 10^{34} joules or 10^{15} GAEE's could do for the civilization that has at its disposal the total energy output of its sun.

Energy Efficiency of Starflight Propulsion

A word about the efficiency of interstellar propulsion. In the foregoing, we spoke as though energy expenditure was convertible to payload kinetic energy at 100% efficiency. Of course this is not so, but how good was the approximation? Realize that factors of 2 or even 10 multiplying the listed energy requirements for payloads will probably not deter an energy-wealthy spacefaring civilization, but it is of interest to know the comparative efficiency of starflight propulsion systems.

One need not be concerned with propulsion systems for which the energy is free or nearly so. Even though systems like the solar sail or the Bussard interstellar ramjet have computable efficiencies for the conversion of sunlight or interstellar fusion fuel into kinetic energy, these "fuels" are obtained free from the environment at no direct cost to civilization. This makes the solar sail and the ramjet extremely attractive.

Rocket fuel, however, is not free, nor is energy (not derived from the Sun) that must be generated and beamed across space. First consider a *nonrelativistic* treatment of propulsion energy efficiency, ϵ, which will be sufficient for our needs:

The Classical Rocket

One way to define the efficiency for a one-stage rocket in which M_l is payload mass (including vehicle structure that achieves final velocity), M_p is propellant mass, V the final payload velocity, and V_e the exhaust jet velocity:

$$\epsilon = \frac{\frac{1}{2}M_l V^2}{\frac{1}{2}M_p V_e^2}$$

To find e as a function of the ratio V/V_e, employ the one-stage rocket equation:

$$\frac{V}{V_e} = \ln\left(\frac{M_p + M_l}{M_l}\right)$$

Substituting this in the previous equation, get:

$$\epsilon = \frac{(V/V_e)^2}{(e^{V/V_e} - 1)}$$

253

Maximizing ϵ with respect to V/V_e by differentiating, get:

$$\epsilon_{max} = 0.647$$

when $V/V_e = 1.60$.

It turns out that the maximum efficiency for a *multistage* rocket occurs at a V/V_e ratio below about 1.60, depending on the structural factors for each stage. Energy efficiency increases from a value lower than 0.647 with increasing number of stages, but doesn't rises above this value, even for very low structural factors.

Beamed Power Rocket

The energy efficiency of a beamed power *rocket* is, of course, the energy efficiency of the classical rocket multiplied by a factor that represents energy conversion efficiency from the original source of energy to exhaust energy:

$$\epsilon = f\epsilon_{rocket}$$

where

$$f = \frac{\text{Directed kinetic energy of exhaust}}{\text{Energy consumption of source feeding the beam}}$$

Energy Efficiency in the Relativistic Regime

G. Marx considered the energy efficiency of propulsion systems in the relativistic flight regime [1,2]. His result for the single-stage classical rocket:

$$\epsilon = \frac{B\left(1 - \sqrt{1 - \beta^2}\right)^{-1}}{\left(1 + B + \sqrt{B(B+2)}\right)^{1/\beta} - 1}$$

where

$$B = \frac{1}{\sqrt{1 - \left(\frac{V}{c}\right)^2}} - 1$$

$$\beta = \frac{V_e}{c}$$

For small V/c, this result reduces, of course, to the nonrelativistic expression derived earlier.

In a more recent study, also incorporating a relativistic treatment, Bernard Oliver discusses ways to further improve rocket efficiency by varying the exhaust velocity during the flight, and he concludes: "In estimating the lower bound of energy needed for interstellar travel we will therefore assume $\epsilon = 1$. The absolute minimum energy needed for a one-way or a round-trip journey in a pre-fueled rocket is the kinetic energy of the payload—i.e., the vehicle left after the last acceleration" (3).

So our conclusion remains valid about the rough order of energy cost for at least one major proposed propulsion class—the rocket. Because of their inherent energy conversion inefficiencies, beamed power propulsion systems surprisingly may be *less* efficient than rockets. But if their energy comes free from the Sun (e.g., a solar-pumped laser), beamed power systems are still to be preferred over the rocket.

Realizing Starflight: A Plan for the Future

Every technological and scientific step undertaken now and in the future, no matter how seemingly trivial, is another mark on the road to practical interstellar travel. In fact, virtually the entire scientific-technological enterprise is already unwittingly contributing to the goal of starflight. For example:

• Discoveries in biology may build the path to significant lifespan extension, hibernation, or even suspended animation.

• Research on controlled thermonuclear power for terrestrial needs may pave the way to working fusion rockets that use magnetic confinement techniques, and discoveries about inertial confinement fusion will lay the groundwork for possible nuclear pulse propulsion systems in the mold of Daedalus.

• Engineering studies and development of solar-power satellites may open the possibility of employing beamed power propulsion for "Starwisp" probes.

• To be sure, studies of nearly "closed" (full material recycling) or completely closed interplanetary space colonies are laying the foundation for interstellar arks or worldships.

• Research on materials for interplanetary solar sails may lead to breakthroughs in materials for interstellar solar sails.

• Basic materials research will promote aspects of all interstellar propulsion concepts.

• Development of high critical temperature superconducting magnets will foster new thinking about interstellar ramjets and ramscoops, electromagnetic launching, and fusion rockets.

• Microminiaturization of electronics and mechanical devices—so-called "nanotechnology"—may lead to ultrasmall and low-mass robotic interstellar probes.

• Research in artificial intelligence and neural networks will make small robotic interstellar probes highly intelligent and increasingly appealing as vehicles for interstellar exploration.

• Development of robotics and fault-tolerant computers will support autonomous interstellar probes.

• The inexorable improvement of both optical and radio astronomy techniques, particularly long-baseline interferometry, may result in discoveries of numerous extrasolar planets in the Solar neighborhood, including Earth-size bodies. Advanced astronomical techniques may be able to reveal environmental conditions on such planets or even detect the atmospheric gas signatures of life processes.

The following new astronomical techniques will also contribute to methods of navigating in interstellar space.

• Astronomical research on the interstellar medium—gases, dust grains, magnetic fields, cosmic radiation, and so on—will, of course, promote starflight through the identification of potential flight hazards, environmental fuel sources, and deceleration/maneuvering technology.

• High-energy and other fundamental physics experiments, as well as additions to fundamental physical theory may reveal unforeseen possibilities for starflight propulsion.

• Above all, terrestrial and later Solar System-wide economic growth, including massive development of extraterrestrial resources, will establish the foundation and infrastructure for starflight. A stable and peaceful global or interplanetary civilization would also encourage the channeling of human energies for the conquest of the interstellar gulf.

Technological horizons are notoriously difficult to foresee, so we will not bore you with those famous bar charts that extend to the year 2500 and beyond and purport to predict the course of space development for all time—with a precision of plus or minus 25 years! Instead, consider the following three eras and what direct actions we might carry out during each of them to realize starflight.

Near-Term (1989–2010)

Astronomy: Establish orbiting "Great Observatories" and use them to determine the frequency of extrasolar planets, particularly Earthlike worlds. Develop and test ground and space-based optical interferometry techniques.

Planetary Science: Inventory the resources of the Solar System, including planetary atmospheres (for fusion fuels), asteroids, and comets.

Life Sciences: Investigate the long-term effects of zero or reduced gravity. Refine estimates of hazards of space radiation to human health. Develop new concepts and more sensitive devices for microbial life detection experiments. Intensify SETI effort nationally and internationally. Support research in cryobiology and hibernation.

Propulsion: Develop and fly prototype solar sails. Conduct space tests of laser-propelled model sails and low-power microwave Starwisp prototypes. Develop mass-drivers and electromagnetic launchers.

Radio Astronomy: Develop long-baseline interferometry for accurate stellar distance determination and astrometry.

Plasma Physics: Investigate mechanisms of erosion by the interstellar medium. Study feasibility of fusion rocketry and antimatter propulsion.

Politics: Strive for recognition of the search for extraterrestrial life and civilizations as a major objective and common goal of global civilization. (We will not hold our breath while we wait on this one!)

Mid-Term (2011–2050)

Establish space colonies and begin widespread utilization of extraterrestrial resources. Investigate social dynamics of space habitats. Launch interstellar precursor missions to 1000 AU and beyond using solar sail or nuclear-electric propulsion. Construct solar-power satellites and launch "symbolic" 0.2 c Starwisp probe.

Far-Term (2051 and Beyond)

Launch robot probes to nearby stars. Base starship designs on experience with Solar-System space colonies and plan the first interstellar colonization mission to depart before the year 2200.

References

CHAPTER 1

1. E. M. Purcell, "Radioastronomy and Communication through Space," *Brookhaven National Laboratory Lectures in Science: Vistas in Research* 1 (November 1960): 3–13 (USAEC Report BNL-658). Also in A. G. W. Cameron, ed., *Interstellar Communication* (New York: W.A. Benjamin, 1963), 121–143.
2. A. R. Martin, ed., *Project Daedalus: The Final Report on the BIS Starship Study*, supplement to *JBIS*, 1978.
3. L. D. Jaffe et al., "An Interstellar Precursor Mission," *JBIS* 33, (January 1980): 3–26 (JPL Publication 77-70, Jet Propulsion Lab, Pasadena, CA).
4. Brice N. Cassenti, "A Comparison of Interstellar Propulsion Methods," *JBIS* 35 (March 1982): 116–124.
5. J. D. Bernal, *The World, The Flesh, and the Devil* (1929; reprint, Bloomington, IN: Indiana University Press, 1969).
6. Richard D. Johnson and Charles Holbrow, eds., *Space Settlements: A Design Study*, NASA SP-413, 1977.
7. Gerard K. O'Neill, *The High Frontier: Human Colonies in Space* (New York: William Morrow, 1977).
8. Robert L. Forward, "Feasibility of Interstellar Travel," *JBIS* 39 (September 1986): 379–384.
9. R. J. Cesarone, A. B. Sergeyevsky, and S. J. Kerridge, "Prospects for the Voyager Extra-Planetary and Interstellar Mission," *JBIS* 37 (March 1984): 99–116.

CHAPTER 2

1. F. H. Crick and Leslie E. Orgel, "Directed Panspermia," *Icarus* 19 (July 1973): 341–346.
2. M. Meot-ner and Gregory L. Matloff, "Directed Panspermia: A Technical and Ethical Evaluation of Seeding Nearby Solar Systems," *JBIS* 32 (November 1979): 419–423.
3. Francis Crick, *Life Itself: Its Origin and Nature* (New York: Simon and Schuster, 1981).
4. B. Zuckerman, "Space Telescopes, Interstellar Probes and Directed Panspermia," *JBIS* 34 (9): 367–370.
5. Robert A. Freitas, Jr., "The Case for Interstellar Probes," *JBIS* 36 (November 1983): 490–495.
6. Robert A. Freitas, Jr., "Interstellar Probes: A New Approach to SETI," *JBIS* 33 (March 1980): 95–100.
7. Ben R. Finney and Eric M. Jones, *Interstellar Migration and the Human Experience* (Berkeley: University of California Press, 1985).

8. L. D. Jaffe et al., "An Interstellar Precursor Mission," *JBIS* 33 (January 1980): 3–26 (JPL Publication 77-70, Jet Propulsion Lab, Pasadena, CA).
9. L. D. Jaffe and H. N. Norton, "A Prelude to Interstellar Flight," *Astronautics and Aeronautics* 18 (January 1980): 38–44.
10. C. W. Allen, *Astrophysical Quantities*, 3rd ed. (London: Athlone Press, 1973).
11. Peter van de Kamp, "Stars Nearer than Five Parsecs," *Sky and Telescope* 14 (October 1955): 498–99.
12. H. R. Mattinson, "Project Daedalus: Astronomical Data on Nearby Stellar Systems," *JBIS* 29 (February 1976): 76–93. Also in *Project Daedalus: The Final Report on the BIS Starship Study*, supplement to *JBIS*, 1978, 8–18.
13. L. W. Alvarez, W. Alvarez, F. Asaro, and H. V. Michel, "Extraterrestrial Cause for the Cretaceous-Tertiary Extinction," *Science* 208 (June 1980): 1095–1108.
14. Richard Muller, *Nemesis: The Death Star* (New York: Weidenfeld & Nicolson, 1988).
15. Marc Davis, Piet Hut, and Richard A. Muller, "Extinction of Species by Periodic Comet Showers," *Nature* 308 (April 1984): 715–717.
16. Walter Alvarez and Richard A. Muller, "Evidence from Crater Ages for Periodic Impacts on Earth," *Nature* 308 (April 1984): 718–720.
17. L. G. Despain, J. P. Hennes, and J. L. Archer, "Scientific Goals of Missions Beyond the Solar System," in *Advances in the Astronautical Sciences*, vol. 29, part 2, *The Outer Solar System*, ed. J. Vagners (Tarzana, CA: American Astronautical Society, 1971; AAS 71-163), 597–616.
18. J. L. Archer and A. J. O'Donnell, "The Scientific Exploration of Near Stellar Systems," in *Advances in the Astronautical Sciences*, vol. 29, part 2, *The Outer Solar System*, ed. J. Vagners (Tarzana, CA: American Astronautical Society, 1971; AAS 71-166), 691–724.
19. A. R. Martin, "Project Daedalus: The Ranking of Nearby Stellar Systems for Exploration," *Project Daedalus: The Final Report on the BIS Starship Study*, supplement to *JBIS*, 1978, 33–36.
20. R. Brandenberg, "A Survey of Interstellar Missions," Technical Memorandum M-30, IIT Research Institute, Chicago, IL, July 1971.
21. Freeman, Dyson, *Infinite in All Directions* (New York: Harper & Row, 1988), 287.

22. Fred Hoyle, *The Intelligent Universe* (New York: Holt, Rinehart and Winston, 1983), 155.
23. John P. Wiley, Jr., *Natural History*, January 1970, 73.
24. Patrick Moore, *The Next Fifty Years in Space* (New York: Taplinger, 1976), 129.
25. Werner von Braun, "Can We Ever Go to the Stars?," *Popular Science*, July 1963, 170.
26. Werner von Braun, correspondence, 4 December 1969.

CHAPTER 3

General Propulsion

1. George P. Sutton, *Rocket Propulsion Elements: An Introduction to the Engineering of Rockets*, 3rd ed. (New York: John Wiley & Sons, 1963).
2. Philip G. Hill and Carl R. Peterson, *Mechanics and Thermodynamics of Propulsion* (Reading, MA: Addison-Wesley, 1965.)
3. Donald L. Turcotte, *Space Propulsion* (New York: Blaisdell Publishing, 1965).
4. Arthur I. Berman, *The Physical Principles of Astronautics* (New York: John Wiley & Sons, 1961).
5. Harry O. Ruppe, *Introduction to Astronautics*, vols. 1 and 2 (New York: Academic Press, 1966).

Electric Propulsion

6. Robert G. Jahn, *Physics of Electric Propulsion* (New York: McGraw-Hill, 1968).
7. Graeme Aston, "Electric Propulsion: A Far Reaching Technology," *JBIS* 39 (November 1986): 503–507.
8. G. L. Matloff, "Electric Propulsion and Interstellar Flight," AIAA Paper 87-1052, 19th AIAA/DGLR/JSASS International Electric Propulsion Conference, May 11–13, 1987, Colorado Springs, CO.

Fission Nuclear Propulsion

9. R. W. Bussard and R. D. DeLauer, *Fundamentals of Nuclear Flight* (New York: McGraw-Hill, 1965).
10. Giovani Vulpetti, "Direct Fission Propulsion: Improvement of a Series-Staged Starship from Impulsive Jettisoning Policy," *JBIS* 31 (March 1978): 93–102.
11. Giovani Vulpetti and I. Mazzitelli, "Probe Fission-Drive: Dynamics and Payload Optimization," *JBIS* 28 (August 1975): 563–572.
12. Giovani Vulpetti, "A Starting Model for a Fission Engine," *JBIS* 28 (August 1975): 573–578.

Fusion Rockets

13. M. U. Clauser, "The Feasibility of Thermonuclear Propulsion," in Proc. of the Conference on Extremely High Temperatures, Air Force Cambridge Research Center, March 18–19, 1958.
14. M. U. Clauser, "Application of Thermonuclear Reactions to Rocket Propulsion," chap. 18, sec. 10 in *Space Technology*, ed. H. S. Seifert (New York: John Wiley & Sons, 1959).
15. R. F. Cooper and R. L. Verga, "Controlled Thermonuclear Reactions for Space Propulsion," ASD-TDR-63-696, Air Force Systems Command, Aeronautical Systems Division, Wright Patterson AFB, Ohio, September 1963.
16. D. Dooling, "Controlled Thermonuclear Fusion for Space Propulsion," *Spaceflight* 14 (January 1972): 26–27.
17. G. W. Englert, "Study of Thermonuclear Propulsion Using Superconducting Magnets," *Engineering Aspects of Magnetohydrodynamics*, ed. N. W. Mather and G. W. Sutton (New York: Gordon and Breach, 1961), 645–671.
18. G. W. Englert, "Towards Thermonuclear Rocket Propulsion," *New Scientist* 4 October 1962 (no. 307), 16–18.
19. J. L. Hilton, J. S. Luce, and A. S. Thompson, "Hypothetical Fusion Propulsion Rocket Vehicle," *Journal of Spacecraft and Rockets* 1 (May-June 1964): 276–282.
20. W. I. Linlor and M. U. Clauser, "Fusion Plasma Propulsion System," in *Advanced Propulsion Concepts*, vol. 1 (New York: Gordon and Breach, 1963), 381–407 (from Proc. 3rd Symposium on Advanced Propulsion Concepts, October 1962, Cincinnati, Ohio).
21. J. S. Luce, "Controlled Fusion Propulsion," in *Advanced Propulsion Concepts*, vol. 1 (New York: Gordon and Breach, 1963), 343-380 (from Proc. 3rd Symposium on Advanced Propulsion Concepts, October 1962, Cincinnati, Ohio).
22. S. H. Masten, "Fusion for Space Propulsion," *IRE Transactions on Military Electronics*, vol. MIL-3, no. 2 (April 1959): 52–57.
23. G. L. Matloff and H. H. Chiu, "Some Aspects of Thermonuclear Propulsion," *J. Astronautical Sciences* 18 (July-August 1970): 57–62.
24. T. C. Powell, "Fusion Power for Interstellar Flight" (Ph.D. diss., Kentucky University, Lexington, Kentucky, 1970); available through NTIS (N71-38275).
25. C. Powell, O. J. Hahn, and J. R. McNally, Jr., "Energy Balance in Fusion Rockets," *Astronautica Acta* 18 (February 1973): 59–69.
26. J. J. Reinmann, "Fusion Rocket Concepts," NASA TM X-67826 (from 6th Symposium on Advanced Propulsion Concepts, Niagara Falls, New York, May 4–6, 1971).
27. J. R. Roth, "A Preliminary Study of Thermonuclear Rocket Propulsion," *JBIS* 18 (1961–1962): 99–108.
28. J. R. Roth, W. Rayle, and J. J. Reinmann, "Fusion Power for Space Propulsion," *New Scientist*, 20 April 1972 (vol. 54), 125–127.

29. D. G. Samaras, "Thermodynamic Considerations of Thermonuclear Space Propulsion," in *Proc. of the 16th International Astronautical Congress (Propulsion and Re-Entry), Athens, 1965* (New York: Gordon and Breach, 1966), 305–322.

30. E. Sänger, "Stationary Nuclear Burning in Rockets" (in German) *Astronautica Acta* 1 (1955): 63–88, 1955; English translation in NASA TM-1405 (1957).

31. E. Sänger, "Pure Fusion Rockets," in *Space Flight* (New York: McGraw-Hill, 1965), 241–255.

32. D. F. Spencer, "Fusion Propulsion System Requirements for an Interstellar Probe," Jet Propulsion Laboratory Technical Report No. 32-397, 15 May 1963.

33. D. F. Spencer, "Fusion Propulsion for Interstellar Missions," in *Annals of the New York Academy of Sciences* 140 (16 December 1966): 407–418.

Antimatter Propulsion

34. E. Sänger, "The Theory of Photon Rockets," in *Space Flight Problems*, (Biel-Bienne, Switz.: Laubscher, 1953), 32–40 (from lecture at 4th International Astronautical Congress).

35. P. F. Massier, "The Need for Expanded Exploration of Matter-Antimatter Annihilation for Propulsion Applications," *JBIS* 35 (September 1982): 387–390.

36. Robert L. Forward, "Antimatter Propulsion," *JBIS* 35 (September 1982): 391–395.

37. Robert L. Forward and Joel Davis, *Mirror Matter: Pioneering Antimatter Physics* (New York: John Wiley & Sons, 1988).

38. B. N. Cassenti, "Design Considerations for Relativistic Antimatter Rockets," *JBIS* 35 (September 1982): 396–404.

39. D. L. Morgan, "Concepts for the Design of an Antimatter Annihilation Rocket," *JBIS* 35 (September 1982): 405–413.

40. R. R. Zito, "The Cryogenic Confinement of Antiprotons for Space Propulsion Systems," *JBIS* 35 (September 1982) : 414–422.

41. G. Chapline, "Antimatter Breeders?" *JBIS* 35 (September 1982): 423–424.

42. Richard R. Zito, "Antimatter Reactor Dynamics," *JBIS* 39 (March 1986): 110–113.

43. Giovani Vulpetti, "An Approach to the Modelling of Matter-Antimatter Propulsion Systems," *JBIS* 37 (September 1984): 403–409.

Relativistic Rockets: Kinematics and Dynamics

44. J. Ackeret, "On the Theory of Rockets," *JBIS* 6 (March 1947): 116–123.

45. S. von Hoerner, "The General Limits of Space Travel," in *Interstellar Communication*, ed. A. G. W. Cameron (New York: W. A. Benjamin, 1963), 144–159. Also in *Science*, 6 July 1962 (vol. 137), 18–23, and *IEEE Student Journal* 1 (March 1963): 21–27.

46. E. Sänger, "On the Attainability of the Fixed Stars" (in German), in *Proc. 7th International Astronautical Congress, Rome, 17–22 September, 1956*, 89–133.

47. L. R. Shepherd, "Interstellar Flight," *JBIS* 11 (1952): 149ff. Also in *Realities of Space Travel*, ed. L. J. Carter (New York: McGraw-Hill, 1957), 395–416.

48. R. Brandenburg, "A Survey of Interstellar Missions," IIT Research Institute Technical Memorandum M-30, July, 1971.

49. Dwain F. Spencer and Leonard D. Jaffe, "Feasibility of Interstellar Travel," Jet Propulsion Laboratory Technical Report No. 32-233, March 15, 1962.

50. Gerald M. Anderson and Donald T. Greenwood, "Relativistic Rocket Flight with Constant Acceleration," *AIAA Journal* 7 (February 1969): 343–344.

51. Gerald M. Anderson and Donald T. Greenwood, "Relativistic Flight with a Constant Thrust Rocket," *Astronautica Acta* 16 (1971): 153–158.

52. John Huth, "Relativistic Theory of Rocket Flight with Advanced Propulsion Systems," *ARS Journal*, March 1960, 250–253.

53. J. M. J. Kooy, "On Relativistic Rocket Mechanics," *Astronautica Acta* 4 (1958): 31–58.

54. Ernst Stuhlinger, "Relativistic Rocket Mechanics," in *Ion Propulsion for Spaceflight*, (New York: McGraw-Hill, 1964), 154–167.

55. E. Sänger, "Some Optical and Kinematical Effects in Interstellar Astronautics," *JBIS* 18 (1961-1962): 273–277.

Miscellaneous

56. Leik Myrabo and Dean Ing, *The Future of Flight* (New York: Baen Enterprises, 1985).

57. P. W. Garrison, R. H. Frisbee, and M. F. Pompa, "Ultra High Performance for Planetary Spacecraft," *FY '81 Final Report*, NASA Jet Propulsion Laboratory, January 1982.

CHAPTER 4

1. A. R. Martin and A. Bond, "Nuclear Pulse Propulsion: An Historical Review of an Advanced propulsion Concept," *JBIS* 32 (August 1979): 283–310.

2. John McPhee, *The Curve of Binding Energy* (New York: Farrar, Strauss and Giroux, 1974).

3. T. W. Reynolds, "Effective Specific Impulse of External Nuclear Pulse Propulsion Systems," *J. Spacecraft and Rockets* 10 (October 1973): 629–630. Also in NASA TN D-6984, September 1972.

4. Freeman J. Dyson, "Death of a Project," *Science*, 9 July 1965 (vol. 149), 141–144.

5. Freeman J. Dyson, "Interstellar Transport," *Physics Today*, October 1968, 41–45.

6. A. Bond, and A. R. Martin, "The Propulsion System," parts 1 and 2, *Project Daedalus: The Final Report on the BIS Starship Study*, supplement to *JBIS*, 1978, 44–82.

7. K. Boyer and J. D. Balcomb, "Systems Studies of Fusion powered Pulsed Propulsion Systems," AIAA Paper No. 71-636, June 14–18, 1971.

8. R. Hyde, L. Wood, and J. Nuckolls, "Prospects for Rocket Propulsion with Laser Induced Fusion Microexplosions," AIAA Paper No. 72-1063, December 1972.

9. F. Winterberg, "Rocket Propulsion by Staged Thermonuclear Microexplosions," *JBIS* 30 (September 1977): 333–340.

10. F. Winterberg, "Rocket Propulsion by Nuclear Microexplosions and the Interstellar Paradox," *JBIS* 32 (November 1979): 403–409.

11. W. A. Reupke, "Inertial Fusion Systems Studies and Nuclear Pulse Propulsion," *JBIS* 38 (November 1985): 483–493.

12. A. P. Fraas, "The Blascon: An Exploding Pellet Fusion Reactor," Oak Ridge National Laboratory TM-3231 (1971).

13. A. R. Martin, ed., *Project Daedalus: The Final Report on the BIS Starship Study*, supplement to *JBIS*, 1978.

14. C. J. Everett and S. M. Ulam, "On a Method of Propulsion of Projectiles by Means of External Nuclear Explosions," Los Alamos Report LAMS-1955 (declassified 25 August 1976).

15. J. C. Nance, "Nuclear Pulse Propulsion," *IEEE Transactions on Nuclear Science* vol. NS-18 (February 1965): 177–182.

16. T. B. Taylor, "Propulsion of Space Vehicles," in *Perspectives in Modern Physics*, ed. R. E. Marshak (New York: Interscience, 1966), 625–640.

17. Kenneth Brower, *The Starship and the Canoe* (New York: Holt, Rinehart and Winston, 1978).

CHAPTER 5

1. R. L. Forward, "Pluto: The Gateway to the Stars," *Missiles and Rockets* 10 (2 April 1962): 26–28.

2. R. L. Forward, "A Programme for Interstellar Exploration," *JBIS* 29 (1976): 611–632.

3. R. L. Forward, "Zero Thrust Velocity Vector Control for Interstellar Probes: Lorentz Force Navigation and Circling," *AIAA Journal* 2 (1964): 885–889.

4. P. C. Norem, "Interstellar Travel: A Round Trip Propulsion System with Relativistic Velocity Capabilities," American Astronautical Society Paper AAS 69-388.

5. R. L. Forward, "Roundtrip Interstellar Travel Using Laser-Pushed Lightsails," *Journal of Spacecraft and Rockets* 21 (1984): 187–195.

6. G. Marx, "Interstellar Vehicle Propelled by Terrestrial Laser Beam," *Nature* 211 (1966): 22–23.

7. J. L. Redding, "Interstellar Vehicle Propelled by Terrestrial Laser Beam," *Nature* 213 (1967): 588–589.

8. W. E. Moeckel, "Propulsion by Impinging Laser Beams," *Journal of Spacecraft and Rockets* 9 (1972): 942–944.

9. J. H. Bloomer, "The Alpha Centauri Probe," in *Proceedings of the 17th International Astronautical Congress (Propulsion and Re-Entry)* (Gordon and Breach, 1967), 225–232.

10. F. Dyson, "Interstellar Propulsion Systems," in *Extraterrestrials: Where are They?*, ed. M. H. Hart and B. Zuckerman (New York: Pergamon, 1982), 41–45.

11. K. E. Drexler, "High Performance Solar Sails and Related Reflecting Devices," AIAA Paper 79-1418, pp. 14–17 (from Fourth Princeton/AIAA Conference on Space Manufacturing Facilities, Princeton, NJ, May 1979).

12. R. L. Forward, "Alternate Propulsion Energy Sources," AFRPL 83-039 and AFRPL 83-067, USAF Rocket Propulsion Laboratory. Also, "Advanced Space Propulsion Study: Antiproton and Beamed Power Propulsion," AFAL TR-87-070, USAF Astronautics Laboratory.

13. G. L. Matloff, "Interstellar Solar Sailing: Consideration of Real and Projected Sail Materials," *JBIS* 37 (1984): 135–141.

14. R. L. Forward, "Starwisp: An Ultralight Interstellar Probe," *Journal of Spacecraft and Rockets* 22 (1985): 345–350.

15. G. L. Matloff and E. Mallove, "Solar Sail Starships: The Clipper Ships of the Galaxy," *JBIS* 34 (1981): 371–380.

16. G. L. Matloff, "Faster Non-Nuclear Worldships," *JBIS* 39 (1986): 475–485.

17. E. Jones, "A Manned Interstellar Vessel Using Microwave Propulsion: A Dysonship," *JBIS* 38 (1985): 270–273.

18. J. T. Early, "Space Transportation System with Energy Transfer and Force Beams," *JBIS* 40 (1987): 371–372.

19. L. Niven and J. Pournelle, *The Mote in God's Eye* (New York, Simon and Schuster, 1974).

20. A. R. Kantrowitz, "Propulsion to Orbit by Ground Based Lasers," *Astronautics and Aeronautics* 10 (May 1972): 74–76.

21. L. Myrabo and D. Ing, *The Future of Flight* (New York: Baen, 1985).

22. A. A. Jackson IV, and D. P. Whitmire, "A Laser-Powered Interstellar Rocket," *JBIS* 31 (1978): 335–337.

23. D. P. Whitmire and A. A. Jackson IV, "Laser Powered Interstellar Ramjet," *JBIS* 30 (1977): 223–226.

24. G. L. Matloff, "Beyond the Thousand Year Ark," *JBIS* 36 (1983): 483–489.

25. Graeme Aston, "Electric Propulsion: A Far Reaching Technology," *JBIS* 39 (November 1986): 503–507.

26. G. L. Matloff, "Electric Propulsion and Interstellar

Flight," AIAA Paper 87-1052, 19th AIAA/DGLR/JSASS International Electric Propulsion Conference, May 11–13, 1987, Colorado Springs, CO.

27. G. L. Matloff and E. F. Mallove, "The Laser-Electric Ramjet: A Near Term Interstellar Propulsion Alternative" (from Proc. of the AIAA 24th Joint Propulsion Conference, Boston, July 12, 1988).

28. I. Craford and R. L. Forward, "Correspondence," *JBIS* 39 (1986): 328.

29. G. L. Matloff and C. B. Ubell, "Worldships: Prospects for Non-Nuclear Propulsion and Power Sources," *JBIS* 38 (1983): 201–209.

30. R. L. Forward, "Light Levitated Geostationary Cylindrical Orbits Using Perforated Light Sails," *Journal of the Astronautical Sciences* 32 (1984): 221–226.

CHAPTER 6

1. Konstantin Tsiolkovskii, *Collected Works* (Moscow: Izd. Akademii Nauk U.S.S.R., 1951, 1954, and 1959). English translations in NASA Technical Translations F-236, F-237, and F-238 (all 1965) and in *Works on Rocket Technology*, NASA TT-F-243 (1965).

2. Fridrikh Tsander, *Problems of Flight by Jet Propulsion*, 1932 (Jerusalem: Israel Program for Scientific Translations, 1964).

3. Carl Wiley, "Clipper Ships of Space," *Astounding Science Fiction* May 1951.

4. Richard L. Garwin, "Solar Sailing: A Practical Method of Propulsion within the Solar System," *Jet Propulsion* 28 (March 1958): 188–190.

5. T. C. Tsu, "Interplanetary Travel by Solar Sail," *ARS Journal* 29 (June 1959): 422–427.

6. Arthur C. Clarke, *Boy's Life* (1964).

7. Gregory L. Matloff and Eugene F. Mallove, "Solar Sail Starships: Clipper Ships of the Galaxy," *JBIS* 34 (September 1981): 371–380.

8. Gregory L. Matloff and Eugene F. Mallove, "The Interstellar Solar Sail: Optimization and Further Analyses," *JBIS* 36 (May 1983): 201–209.

9. Gregory L. Matloff and Eugene F. Mallove, "The First Interstellar Colonization Mission," *JBIS* 33 (March 1980): 84–88.

10. Gregory L. Matloff, "Beyond the Thousand-Year Ark: Further Study of Non-Nuclear Interstellar Flight," *JBIS* 36 (November 1983): 483–489.

11. K. Eric Drexler, "High Performance Solar Sails and Related Reflecting Devices," AIAA Paper No. 79-1418; "High Performance Solar Sail Concept," *L-5 News* 4 (May 1979): 7–9.

12. Gregory L. Matloff, "Interstellar Solar Sailing: Consideration of Real and Projected Sail Material," *JBIS* 37 (March 1984): 135–141.

13. K. A. Ehricke, "Saturn-Jupiter Rebound: A Method of High Speed Spacecraft Ejection from the Solar System," *JBIS* 25 (1972): 561–571.

14. Robert L. Forward, "Infrastellar and Interstellar Exploration," Hughes Research Report 503 (September 1976).

15. V. R. Eshleman, "Gravitational Lens of the Sun: Its Potential for Observations and Communications Over Interstellar Distances," *Science* 205 (1979): 1133–1135.

16. Louis Friedman, *Starsailing: Solar Sails and Interstellar Travel* (New York: John Wiley & Sons, 1988).

17. K. Eric Drexler, "Sailing on Sunlight May Give Space Travel a Second Wind," *Smithsonian* 12 (February 1982): 52.

18. Gregory L. Matloff and Charles B. Ubell, "World Ships: Prospects for Non-Nuclear Propulsion and Power Sources," *JBIS* 38 (June 1985): 253–261.

19. Bernard W. Powell, "Solar Sail: Key to Interplanetary Voyaging?" *Spaceflight* 2 (October 1959): 16–118.

20. Gregory L. Matloff, "The State of the Art Solar Sail and the Interstellar Precursor Mission," *JBIS* 37 (November 1984): 491–494.

21. Gregory L. Matloff, "World Ships and White Dwarfs," *JBIS* 39 (March 1986): 114–115.

CHAPTER 7

1. R. W. Bussard, "Galactic Matter and Interstellar Spaceflight," *Astronautica Acta* 6 (1960): 179–194.

2. W. Sullivan, *We Are Not Alone* (New York: McGraw-Hill, 1964).

3. S. H. Dole and I. Asimov, *Planets for Man* (New York: Random House, 1964).

4. C. Sagan, "Direct Contact Among Galactic Civilizations by Relativistic Spaceflight," *Planetary and Space Science* 11 (1963): 485–498.

5. I. S. Shklovskii and C. Sagan, *Intelligent Life in the Universe* (New York: Dell, 1966).

6. L. Niven, *A Gift From Earth* (New York: Ballantine, 1968).

7. P. Anderson, *Tau Zero* (Garden City, NY: Doubleday, 1970).

8. J. F. Fishback, "Relativistic Interstellar Spaceflight," *Astronautica Acta* 15 (1969): 25–35.

9. A. R. Martin, "Structural Limitations on Interstellar Spaceflight," *Astronautica Acta* 16 (1971): 353–357.

10. A. R. Martin, "Magnetic Intake Limitations on Interstellar Ramjets," *Astronautica Acta* 18 (1973): 1–10.

11. F. Dyson, "Interstellar Propulsion Systems," in *Extraterrestrials: Where Are They?*, ed. M. H. Hart and B. Zuckerman (New York: Pergamon, 1982): 41–45.

12. T. A. Heppenheimer, "On the Infeasibility of Interstellar Ramjets," *JBIS* 31 (1978): 222–224.

13. R. L. Forward, "Feasibility of Interstellar Travel," *JBIS* 39 (1986): 379–384.

14. D. P. Whitmire, "Relativistic Spaceflight and the Catalytic Nuclear Ramjet," *Acta Astronautica* 2 (1975): 497–509.

15. B. C. Maglich, "The Migma Principle of Controlled Fusion," *Nuclear Instruments and Methods* 111 (1973): 213–235.

16. G. L. Matloff and A. J. Fennelly,"Vacuum-Ultraviolet Laser and Interstellar Flight," *JBIS* 28 (1975): 443

17. B. Karlovitz and B. Lewis, "Space Propulsion by Interstellar Gas," in *Proc. of 9th International Astronautical Congress* (New York: Springer-Verlag, 1959), 307–311.

18. A. Bond, "An Analysis of the Potential Performance of the Ram Augmented Interstellar Rocket," *JBIS* 27 (1974): 674–685.

19. C. Powell, "Flight Dynamics of the Ram-Augmented Interstellar Rocket," *JBIS* 28 (1975): 553–562.

20. C. Powell, "System Optimization for the Ram-Augmented Interstellar Rocket," *JBIS* 29 (1976): 136–142.

21. C. Powell, "The Effect of Subsystem Inefficiencies upon the Performance of the Ram-Augmented Interstellar Rocket," *JBIS* 29 (1976): 786–794.

22. A. A. Jackson IV, "Some Considerations on the Antimatter and Fusion Ram Augmented Interstellar Rocket," *JBIS* 33 (1980): 117–120.

23. G. L. Matloff and E. F. Mallove, "The First Interstellar Colonization Mission," *JBIS* 33 (1980): 84–88.

24. A. Dalgarno R. A. McCray, "Heating and Ionization of HI Regions," *Annual Review of Astronomy and Astrophysics* 10 (1972): 375–426.

25. R. C. Bless and A. D. Code, "Ultraviolet Astronomy," *Annual Review of Astronomy and Astrophysics* 10 (1972): 197–226.

26. E. Novotny, *Introduction to Stellar Atmospheres and Interiors* (New York: Oxford, 1973).

27. G. O. Abell, *Exploration of the Universe*, 3rd ed. (New York: Holt, Rinehart and Winston,1975).

28. A. Martin, ed., *Project Daedalus: The Final Report on the BIS Starship Study*, supplement to *JBIS*, 1978.

29. G. L. Matloff, "Utilization of O'Neill's Model I LaGrange Point Colony as an Interstellar Ark," *JBIS* 29 (1976): 775–785.

CHAPTER 8

1. R. W. Bussard, "Galactic Matter and Interstellar Spaceflight," *Astronautica Acta* 6 (1960): 179–194.

2. C. Sagan, "Direct Contact Among Galactic Civilizations by Relativistic Spaceflight," *Planetary and Space Science* 11 (1963): 485–498.

3. J. F. Fishback, "Relativistic Interstellar Spaceflight," *Astronautica Acta* 15 (1969): 25–35.

4. A. R. Martin, "Some Limitations of the Interstellar Ramjet," *Spaceflight* 14 (1972): 21–25.

5. G. L. Matloff and A. J. Fennelly, "A Superconducting Ion Scoop and Its Application to Interstellar Flight," *JBIS* 27 (1974): 663–673.

6. A. Bond, "An Analysis of the Potential Performance of the Ram-Augmented Interstellar Rocket," *JBIS* 27 (1974): 674–685.

7. D. P. Whitmire, "Relativistic Spaceflight and the Catalytic Nuclear Ramjet," *Acta Astronautica* 2 (1975): 497–509.

8. G. L. Matloff, "Utilization of O'Neill's Model I Lagrange Point Colony as an Interstellar Ark," *JBIS* 29 (1976): 775–785.

9. G. L. Matloff and A. J. Fennelly, "Interstellar Applications and Limitations of Several Electrostatic/Electromagnetic Ion Collection Techniques," *JBIS* 30 (1977): 213–222.

10. N. H. Langton, "The Erosion of Interstellar Drag Screens," *JBIS* 26 (1973): 481–484.

11. C. Powell, "Flight Dynamics of the Ram-Augmented Interstellar Rocket," *JBIS* 28 (1975): 553–562.

12. W. B. Roberts, "The Relativistic Dynamics of a Sub-Light Speed Interstellar Ramjet Probe," *JBIS* 29 (1976): 795–812.

13. A. R. Martin, "The Effects of Drag on Relativistic Spaceflight," *JBIS* 25 (1972): 643–652.

14. F. Dyson, "Interstellar Propulsion Systems," in *Extraterrestrials: Where Are They?*, ed. M. H. Hart and B. Zuckerman (New York: Pergamon, 1982), 41–45.

15. G. L. Matloff and E. F. Mallove, "Solar Sail Starships:The Clipper Ships of the Galaxy,"*JBIS* 34 (1981): 371–380.

16. E. F. Mallove, *The Quickening Universe: Cosmic Evolution and Human Destiny* (New York: St. Martin's Press, 1987).

17. R. D. Johnson, ed., "The Plasma Core Shield," chap. 4 in *Space Settlements: A Design Study*, NASA SP-413, 1977.

18. A. Dalgarno and R. A. McCray, "Heating and Ionization of HI Regions," *Annual Review of Astrophysics* 10 (1972): 375–426.

19. A. R. Martin, ed., *Project Daedalus: The Final Report on the BIS Starship Study*, supplement to *JBIS*, 1978.

CHAPTER 9

1. Stanislaw Ulam, "On the Possibility of Extracting Energy from Gravitational Systems by Navigating Space

Vehicles," Los Alamos Scientific Laboratory Report LAMS-2219, June 19, 1958.

2. Stanislaw Ulam, "On Some Statistical Properties of Dynamical Systems," *Proceedings of the Fourth Berkeley Symposium on Mathematical Statistics and Probability, June 20–July 30, 1960*, 315–320.

3. F. J. Dyson, "Gravitational Machines," in *Interstellar Communication* (New York: W. A. Benjamin, 1963).

4. F. J. Dyson, "The Search for Extraterrestrial Technology," in *Perspectives in Modern Physics*, ed. R. E. Marshak (New York: Interscience Publishers, 1966).

5. K. A. Ehricke, "The Ultraplanetary Probe," American Astronautical Society 17th Annual Meeting, June 28–30, 1971, Preprint No. AAS-71-164. Also in *Advances in the Astronautical Sciences*, vol. 29, part 2, *The Outer Solar System*, ed. J. Vagners (Tarzana, CA: AAS), 617–679.

6. K. A. Ehricke, "Saturn-Jupiter Rebound: A Method of High Speed Spacecraft Ejection from the Solar System," *JBIS* 25 (1972): 561–571.

7. Gregory L. Matloff and Eugene F. Mallove, "Non-Nuclear Interstellar Flight: Application of Planetary Gravity Assists," *JBIS* 36 (1983): 201–209.

8. Gregory L. Matloff and Kelly Parks, "Interstellar Gravity Assist Propulsion," *JBIS* 41 (November 1988): 519–526.

9. A. C. Clarke, "Electromagnetic Launching as a Major Contribution to Space Flight," *JBIS* 9 (November 1950): 261–267.

10. F. Winterberg, "Magnetic Acceleration of a Superconducting Solenoid to Hypervelocities," *Journal of Nuclear Energy* 8 (1966): 541–553.

11. F. Winterberg, "The Electromagnetic Rocket Gun: A Means to Reach Ultrahigh Velocities," *Atomkernenergie-Kerntechnik* 43 (1983): 104–108.

12. E. H. Lemke, "Magnetic Acceleration of Interstellar Probes," *JBIS* 35 (1982): 498–503.

13. C. F. Singer, "Interstellar Propulsion Using a Pellet Stream for Momentum Transfer," *JBIS* 33 (March 1980): 107–116. Also, Princeton Plasma Physics Laboratory Report PPPL-1583, October 1979.

14. C. F. Singer, "Questions Concerning Pellet-Stream Propulsion," *JBIS* 34 (1981): 117–119.

15. James T. Early, "Space Transportation Systems with Energy Transfer and Force Beams," *JBIS* 40 (August 1987): 371–372.

16. Benoit A. Lebon, "Magnetic Shepherding of Orbital Grain Streams," *JBIS* 39 (November 1986): 486–490.

17. E. F. Mallove, "Scissors: A New Way to the Stars," unpublished communication to Robert L. Forward and Gregory L. Matloff, July 1976.

18. Edmond Rostand, *Cyrano de Bergerac*, trans. Brian Hooker (New York: Bantam Books, 1959).

CHAPTER 10

1. M. E. Davies, and B. C. Murray, *The View from Space*, (New York: Columbia University Press, 1971).

2. R. J. Cesarone, A. B. Sergeyersky, and S. J. Kerridge, "Prospects for the Voyager Extra-Planetary and Interstellar Mission," *JBIS* 37 (1984): 99–116.

3. W. Gleise, *Catalogue of Nearby Stars* Veroeffenlichungen des Astronomischen Rechen-Instituts Heidelberg, no. 22 (Karlsruhe: Verlag G. Braun, 1969).

4. R. E. Wilson, *General Catalogue of Stellar Radial Velocities*, Mount Wilson and Palomar Observatories, Carnegie Institute of Washington, Publication 601 (Washington: 1953).

5. R. Muller, *Nemesis the Death Star: The Story of a Scientific Revolution* (New York: Weidenfeld & Nicolson, 1988).

6. F. J. Dyson, "Gravitational Machines," in *Interstellar Communication*, ed. A. G. W. Cameron (New York: W. J. Benjamin, 1963).

7. G. L. Matloff and E. Mallove, "The Interstellar Solar Sail-Optimization and Further Analysis," *JBIS* 36 (1983): 201–209.

8. C. E. Singer, "Interstellar Propulsion Using a Pellet Stream for Momentum Transfer," *JBIS* 33 (1980): 107–116.

9. G. M. Anderson, "Optimal Interstellar Trajectories with Acceleration Limited Relativistic Rockets," *Journal of the Astronautical Sciences* 15 (1968): 312–318.

10. G. M. Anderson, "Optimal Interstellar Trajectories with Thrust-Limited Relativistic Rockets," *Journal of the Astronautical Sciences* 17 (1969): 65–69.

11. G. M. Anderson, "Optimal Interstellar Relativistic Rocket Trajectories with Both Thrust and Acceleration Constraints," *JBIS* 27 (1974): 273–285 (presented at the 24th International Astronautical Congress, Baku, USSR, October 1983.

12. C. Powell, "Optimal Exhaust Velocity Programming for an Energy-Limited, Single-Stage Relativistic Rocket," *JBIS* 27 (1974): 263–266.

13. C. Powell, "Flight-Time Minimization for an Energy-Limited Flyby Star Probe," *JBIS* 27 (1974): 267–272.

14. C. Powell R. P. Mikkilineni, "Optimal Exhaust Velocity Programming for a Single-Stage Constant Power Starship," *JBIS* 30 (1974): 460–462.

15. P. Anderson, *Tau Zero* (Garden City, NY: Doubleday, 1970).

16. R. W. Bussard, "Galactic Matter and Interstellar Space-flight," *Astronautica Acta* 6 (1960): 179–194.

17. C. Sagan, "Direct Contact Among Galactic Civilizations by Interstellar Spaceflight," *Planetary and Space Science* 11 (1963): 485–498.

18. G. Vulpetti, "Relativistic Astrodynamics: Non-Rectilinear Trajectories for Star Exploration Flights," *JBIS* 34 (1981): 477–485.

19. G. Vulpetti, "Starship Flight Optimization: Time Plus Energy Optimization Criterion," *JBIS* 31 (1978): 403–410.

20. G. Vulpetti, "Multiple Propulsion Concept: Theory and Performance," *JBIS* 32 (1979): 209–214.

21. Francis B. Hildebrand, *Methods of Applied Mathematics*, 2nd ed. (Englewood Cliffs, NJ: Prentice-Hall, 1965).

CHAPTER 11

1. William C. Hinds, *Aerosol Technology* (New York: John Wiley & Sons, 1982).

2. J. S. Stodolkiewicz, *General Astrophysics with Elements of Geophysics* (New York: American Elsevier, 1973).

3. B. Stromgrem, "The Physical State of Interstellar Hydrogen," *Astrophysical Journal* 89 (1939): 526–547. Also in *A Sourcebook in Astronomy and Astrophysics*, ed., K. R. Lang and O. Gingerich, (Cambridge: Harvard University Press, 1979).

4. L. Houziaux and H. E. Butler, eds., *Ultraviolet Stellar Spectra and Related Ground-Based Observations*, International Astronomical Union (IAU) Symposium No. 36 (New York: Springer-Verlag, 1970).

5. L. Spitzer, Jr., and E. B. Jenkins, "Ultraviolet Studies of the Interstellar Gas," *Annual Review of Astronomy and Astrophysics* 13 (1975): 133–164.

6. H. C. van de Hulst, "The Solid Particles of Interstellar Space," *Recherches Astronomiques de l'Observatoire d'Utrecht* 11 (part 2, 1949): 1–50. Also in *A Sourcebook in Astronomy and Astrophysics*, ed. K. R. Lang and O. Gingerich (Cambridge: Harvard University Press, 1979).

7. J. M. Greenberg and P. Weber, "Panspermia: A Modern Astrophysical and Biological Approach," in *The Search for Extraterrestrial Life: Recent Developments*, ed. M. D. Papagiannis (Boston: D. Reidel, 1985).

8. G. Verschuur, "Molecules between the Stars," *Mercury* 16 (1987): 66–76.

9. F. Hoyle, *The Intelligent Universe* (London: Michael Joseph, 1983).

10. G. Benford and D. Brin, *Heart of the Comet* (New York: Bantam, 1986).

11. R. E. Davies, A. M. Delluva, and R. H. Koch, "No Valid Evidence Exists for Interstellar Proteins, Bacteria, etc," in *The Search for Extraterrestrial Life: Recent Developments*, ed. M. D. Papagiannis (Boston: D. Reidel, 1985).

12. G. Verschuur, "Something Passing in the Night," *Astronomy* 15, no. 12 (1987): 26–31.

13. V. F. Hess, "Concerning Observations of Penetrating Radiation in Seven Free Balloon Flights" (in German), *Physikalishe Zeitschrift* 13 (1912): 1084–1091; English translation by B. Doyle in *A Sourcebook in Astronomy and Astrophysics*, ed. K. R. Lang and O. Gingerich (Cambridge: Harvard University Press, 1979).

14. R. A. Millikan, and G. H. Cameron, *Physical Review* 28 (1926): 851.

15. W. Baade and F. Zwicky, "On Super-Novae," *Proc. of National Academy of Sciences* 20 (1934): 254–259. Also in *A Sourcebook in Astronomy and Astrophysics*, ed. K. R. Lang and O. Gingerich (Cambridge: Harvard University Press, 1979).

16. E. Fermi, "Galactic Magnetic Fields and the Origin of Cosmic Radiation," *Astrophysical Journal* 119 (1954): 1–6. Also in *A Sourcebook in Astronomy and Astrophysics*, ed. K. R. Lang and O. Gingerich (Cambridge: Harvard University Press, 1979).

17. K. O. Kiepenheuer, "Cosmic Rays as the Source of General Galactic Radio Emission," *Physical Review* 79 (1950): 738–739, and V. I. Ginzburg, "The Nature of Cosmic Radio Emission and the Origin of Cosmic Rays," *Nuovo Cimento Supplement* 3 (1956): 38–48. Both reprinted in *A Sourcebook in Astronomy and Astrophysics*, ed. K. R. Lang and O. Gingerich (Cambridge: Harvard University Press, 1979).

18. A. R. Martin, "The Effects of Drag on Relativistic Spaceflight," *JBIS* 25 (1972): 643–653.

19. E. T. Benedict, "Disintegration Barriers to Extremely High-Speed Space Travel," *Advances in the Astronautical Sciences* 6 (1961): 571–588.

20. N. H. Langton, "The Erosion of Interstellar Drag Screens," *JBIS* 26 (1973): 481–484.

21. C. Powell, "Heating and Drag at Relativistic Speeds," *JBIS* 28 (1975): 546–552.

22. N. H. Langton and W. R. Oliver, "Materials in Interstellar Flight," *JBIS* 30 (1977): 109–111.

23. A. R. Martin, "Project Daedalus: Bombardment by Interstellar Material and Its Effects on the Vehicle," *Project Daedalus: The Final Report on the BIS Starship Study*, supplement to *JBIS*, 1978, S116–S121.

24. V. W. Bowlie, "Cosmic Background Radiation Drag Effects at Relativistic Speeds," *JBIS* 34 (1981): 499–501.

25. I. Crawford and R. L. Forward, "Correspondence," *JBIS* 39 (1986): 328.

26. J. H. Wolfe, "On the Question of Interstellar Travel,"

The Search for Extraterrestrial Life: Recent Developments, ed. M. D. Papagiannis (Boston: D. Reidel, 1985), 449–454.

27. R. D. Johnson, ed., *Space Settlements: A Design Study*, NASA SP-413, 1977.

28. G. L. Matloff, "Cosmic Ray Shielding for Manned Interstellar Arks and Mobile Habitats," *JBIS* 30 (1977): 96–98.

29. P. Birch, "Radiation Shields for Ships and Settlements," *JBIS* 35 (1982): 515–519.

30. R. Silberberg, C. H. Tsao, J. H. Adams, Jr., and J. Letaw, "Radiation Hazards in Space," *Aerospace America* 25 (1987): 38–41.

31. Hannes Alfven, "Memoirs of a Dissident Scientist," *American Scientist* 26 (May-June 1988): 249–251.

CHAPTER 12

1. Richard H. Battin, *Astronautical Guidance* (New York: McGraw-Hill, 1964).

2. J. R. Wertz, "Interstellar Navigation," *Spaceflight* 14 (June 1972): 206–216.

3. Roger W. Sinnott, "The Wandering Stars of Allegheny," *Sky & Telescope* 64 (October 1987): 360–363.

4. B. F. Burk et al., "Cambridge Workshop on Imaging Interferometry: Final Report," March 1987 (from a NASA-sponsored meeting at the American Academy of Arts and Sciences, Cambridge, MA, October 28–30, 1985).

5. R. B. Phillips and J. F. Lestrade, "Compact Non-Thermal Radio Emission from B-Peculiar Stars," *Nature*, 28 July 1988, 329–331.

6. D. G. Hoag and W. Wrigley, "Navigation and Guidance in Interstellar Space," *Proc. of the 24th IAF Congress, Baku, USSR, October 1–13, 1973*, 513–533.

7. A. T. Lawton, "Stardrift: A Navigational System for Relativistic Interstellar Flight," *Spaceflight* 15 (July 1973): 256–260.

8. G. R. Richards, "Project Daedalus: The Navigation Problem," *Project Daedalus: The Final Report on the Bis Starship Study*, supplement to *JBIS*, 1978, S143–148.

9. W. Markowitz, "Time and Space Navigation," in *Air, Space, and Instruments*, ed., S. Lees (New York: McGraw-Hill, 1963), 201–206.

10. E. Sänger, "Some Optical and Kinematical Effects in Interstellar Astronautics," *JBIS* 18 (1961-1962): 273–277.

11. S. M. Rytov, "What an Astronaut Will See and Encounter When Flying at a Speed Approaching That of Light," *American Rocket Society Journal* 31 (Supplement, 1961): 678–681.

12. B. M. Oliver, "The View form the Starship Bridge and Other Observations," *IEEE Spectrum* 1 (January 1964): 86–92.

13. S. Moskowitz and W. P. Devereux, "Trans-Stellar Space Navigation," *AIAA Journal* 6 (June 1968): 1021–1029 (from report of AIAA/JACC Guidance and Control Conference, Seattle, Washington, August 15–17, 1966).

14. S. Moskowitz and W. P. Devereux, "Navigational Aspects of Trans-Stellar Space Flight," in *Advances in Space Science and Technology*, vol. 10, ed. Frederick I. Ordway III (New York: Academic Press, 1970), 75–126.

15. S. Moskowitz, "Visual Aspects of Trans-Stellar Space Flight," *Sky and Telescope* 33 (May 1967): 290–294.

16. R. W. Stimets and E. Sheldon, "The Celestial View from a Relativistic Starship," *JBIS* 34 (March 1981): 83–99.

17. E. Sheldon and R. H. Giles, "Celestial Views from Non-Relativistic and Relativistic Interstellar Spacecraft," *JBIS* 36 (March 1983): 99–114.

18. J. M. McKinley and P. Doherty, "In Search of the 'Starbow': The Appearance of the Starfield from a Relativistic Spaceship," *American Journal of Physics* 47 (April 1979): 309–316.

19. G. D. Scott and M. R. Viner, "The Geometrical Appearance of Large Objects Moving at Relativistic Speeds," *American Journal of Physics* 33 (1965): 534–536.

20. G. D. Scott and H. J. Van Driel, "Geometrical Appearances at Relativistic Speeds," *American Journal of Physics* 38 (August 1970): 971–977.

21. V. F. Weisskopf, "The Visual Appearance of Rapidly Moving Objects," *Physics Today*, September 1960, 24–27.

22. G. Vulpetti, "A Problem of Relativistic Navigation: The Three-Dimensional Rocket Equation," *JBIS* 31 (September 1978): 344–351.

CHAPTER 13

1. R. L. Forward, "Far Out Physics," *Analog Science Fiction/Science Fact*, August 1975, 146–166.

2. R. L. Forward, *Future Magic* (New York: Avon Books, 1988).

3. J. P. Hogan, *The Gentle Giants of Ganymede* (New York: Ballantine, 1978).

4. N. S. Kardashev et al., "Astroengineering Activity: The Possibility of ETI in Present Astrophysical Phenomena," *Communication with Extraterrestrial Intelligence: CETI*, ed. Carl Sagan (Cambridge: MIT Press, 1973).

5. C. Sagan, *The Cosmic Connection* (Garden City, NY: Doubleday, 1973).

6. A. Berry, *The Iron Sun* (New York: Warner Books, 1977).
7. T. Gold, Discussion of Kardashev's paper at Moscow SETI conference, in *Communication with Extraterrestrial Intelligence: CETI*, ed. Carl Sagan (Cambridge: MIT Press, 1973).
8. G. H. Stine, "Detesters, Phasers, and Dean Drives," *Analog Science Fiction/Science Fact*, June 1976, 62–80.
9. A. Beiser *Basic Concepts of Physics* 2nd ed. (Reading, MA: Addison-Wesley, 1972).
10. P. Birch "Is Faster than Light Travel Causally Possible?," *JBIS* 37 (1984): 117–123.
11. H. D. Froning, "Requirements for Rapid Transport to the Further Stars," *JBIS* 36 (1983): 227–230.
12. R. T. Jones, "Relativistic Kinematics for Motions Faster Than Light," *JBIS* 35 (1982): 509–514.
13. H. D. Froning, "Propulsion Requirements for a Quantum Interstellar Ramjet," *JBIS* 33 (July 1980): 265–270.
14. H. D. Froning, "Use of Vacuum Energies for Interstellar Space Flight," *JBIS* 39 (1986): 410–415.
15. E. F. Mallove, "The Self-Reproducing Universe," *Sky & Telescope* 76 (September 1988): 253–256.
16. J. M. J. Kooy, "Gravitation and Space Flight," *Acta Astronautica* 4 (1977): 229–230.
17. L. Niven, "The Warriors," in *Tales of Known Space* (New York: Ballantine, 1975).
18. F. J. Tipler, "Rotating Cylinders and the Possibility of Global Causality Violation," *Physical Review D* 9 (15 April 1974): 2203–2206.
19. W. O. Davis, "The Fourth Law of Motion," *Analog Science Fact/Science Fiction*, May 1962, 83–104.
20. Michael S. Morris, Kip S. Thorne, and Ulvi Yurtsever, "Wormholes, Time Machines, and Weak Energy Condition," *Physical Review Letters* 61 (26 September 1988): 1446–1449.

CHAPTER 14

1. Frank Drake, "On Hands and Knees in Search of Elysium," *Technology Review* 78 (June 1976): 22–29.
2. John Hands, "Suspended Animation for Space Flight," *JBIS* 38 (March 1985): 139–142.
3. Robert C. W. Ettinger, *The Prospect of Immortality* (Garden City, NY: Doubleday, 1964).
4. H. C. Heller, X. J. Musacchia, and L. C. H. Wang, ed., *Living in the Cold: Physiological and Biochemical Adaptations* (New York: Elsevier, 1986).
5. X. J. Musacchia, "Comparative Physiological and Biochemical Aspects of Hypothermia as a Model for Hibernation," *Cryobiology* 21 (1984): 583–592.
6. Felix Franks, *Biophysics and Biochemistry at Low Temperatures* (Cambridge: Cambridge University Press, 1985).

CHAPTER 15

1. J. L. Archer and A. J. O'Donnell, "The Scientific Exploration of Near Stellar Systems," Proceedings of the AAS 17th Annual Meeting, June 28–30, 1971, Preprint No. AAS-71-166. Also in *Advances in the Astronautical Sciences*, vol. 29, part 2, *The Outer Solar System*, ed. J. Vagners (Tarzana, CA: AAS), 691–724.
2. L. G. Despain, J. P. Hennes, and J. L. Archer, "Scientific Goals of Missions Beyond the Solar System," Proceedings of the AAS 17th Annual Meeting, June 28–30 1971, Preprint No. AAS-71-163. Also in *Advances in the Astronautical Sciences*, vol. 29, part 2, *The Outer Solar System*, ed. J. Vagners (Tarzana, CA: AAS), 597–616.
3. R. Brandenburg, "A Survey of Interstellar Missions," Illinois Institute of Technology Research Institute, Technical Memorandum M-30, July 1971.
4. K. A. Ehricke, "Evolution of Interstellar Operations," American Astronautical Society Joint National Meeting, Denver, CO, June 1969, AAS Paper No.69-387-1, 2.
5. K. A. Ehricke, "Interstellar Mission Concepts," in "The Exploration of the Solar System and Interstellar Space," *Annals of the New York Academy of Sciences* 163:538–553.
6. K. A. Ehrike, "The Ultraplanetary Probe," AAS 17th Annual Meeting, Preprint No. AAS-71-164, June 28–30, 1971. Also in *Advances in the Astronautical Sciences*, vol. 29, part 2, *The Outer Solar System*, ed. J. Vagners (Tarzana, CA: AAS), 617–679.
7. L. D. Jaffe and C. V. Ivie, "Science Aspects of a Mission Beyond the Planets," *Icarus* 39 (1979): 486–494.
8. A. R. Martin, "Project Daedalus: The Ranking of Nearby Stellar Systems for Exploration," *Project Daedalus: The Final Report on the Bis Starship Study*, supplement to *JBIS*, 1978, 33–36.
9. L. D. Jaffe et al., "An Interstellar Precursor Mission," JPL Publication 77–70, Jet Propulsion Lab, Pasadena, CA; *JBIS* 33 (January 1980): 3–26.
10. K. J. Rooney, "Interstellar Probes: A Communications Philosophy," *JBIS* 31 (September 1978): 323–334.
11. Gerald M. Anderson, "Some Problems in Communications with Relativistic Interstellar Rockets," *JBIS* 28 (March 1975): 168–174.
12. T. J. Grant, "Daedalus Probe Requirements and Deployment as a Function of Reliability," *JBIS* 35 (May 1982): 226–234.
13. G. R. Richards, "Planetary Detection from an Interstellar Probe," *JBIS* 28 (August 1975): 579–585.

CHAPTER 16

1. J. M. Pasachoff, *Contemporary Astronomy* (Philadelphia: Saunders, 1977).

2. J. C. Tarter, D. C. Black, and J. Billingham, "Review of Methodology and Technology Available for the Detection of Extrasolar Planetary Systems," *JBIS* 39 (1986): 418–424 (from 36th International Astronautical Congress, Stockholm, Sweden, October 1985).

3. N. G. Roman, "Planets of Other Stars," *Astronomical Journal* 64 (1959): 344–345.

4. L. Spitzer, Jr., "The Beginnings and Future of Space Astronomy," *American Scientist* 50 (1962): 473–484.

5. A. J. Fennelly, G. L. Matloff, and G. Frye, "Photometric Detection of Extrasolar Planets Using L.S.T.–Type Telescopes," *JBIS* 28 (1975): 399–404.

6. J. E. Elliot, "Direct Imaging of Extra-Solar Planets with Stationary Occultations Viewed by a Space Telescope," *Icarus* 35 (1978): 156–163.

7. G. K. O'Neill, "A High Resolution Orbiting Telescope," *Science* 160 (1968): 843–847.

8. S. S. Huang, "Extrasolar Planetary Systems," *Icarus* 18 (1973): 339–376.

9. A. R. Martin, "The Detection of Extrasolar Planetary Systems, parts 1, 2, and 3," *JBIS* 27 (1974): 643–659, 881–906, and *JBIS* 28 (1975): 182–190.

10. G. L. Matloff and A. J. Fennelly, "Optical Techniques for the Detection of Extrasolar Planets: A Critical Review," *JBIS* 29 (1976): 471–481.

11. C. E. Kenknight, "Methods of Detecting Extrasolar Planets: Imaging," *Icarus* 30 (1977): 422–433.

12. W. A. Baum, "The Search for Planets in Other Solar Systems Through Use of the Space Telescope," in *Strategies for the Search for Life in the Universe*, ed. M. D. Papagiannis (Boston: D. Reidel, 1980).

13. D. C. Black, "A Review of the Scientific Rationale and Methods Used in the Search for Other Planetary Systems," in *The Search for Extraterrestrial Life: Recent Developments* I. A. U. Symposium No. 112, ed. M. D. Papagiannis (Boston: D. Reidel, 1985). Also, D. C. Black, "A Comparison of Alternative Methods for Detecting Other Planetary Systems," in *Strategies for the Search for Life in the Universe*, ed. M. D. Papagiannis (Boston: D. Reidel, 1980).

14. J. Greenstein, chairman, Minutes of the First Workshop on Extrasolar Planetary Detection, Lick Observatory, University of California, Santa Cruz, CA, March 23–24, 1976.

15. R. Brown's doubts on planet imaging with the HST are described in *Sky and Telescope* 75 (January 1988).

16. G. Flint, *Sky and Telescope* 71 (May 1984): 402.

17. K. Serkowski, "Search for Planets by Spectroscopic Methods," in *Strategies for the Search for Life in the Universe*, ed. M. D. Papagiannis (Boston: D. Reidel, 1980).

18. W. J. Borucki, J. D. Scargle, and H. S. Hudson, "The Detectability of Extrasolar Transits," *Astrophysical Journal* 291 (1985): 852–854.

19. G. Gatewood, J. Stein, L. Breakiron, R. Goebel, S. Kipp, and J. Russell, "The Astrometric Search for Neighboring Planetary Systems," in *Strategies for the Search for Life in the Universe*, ed. M. D. Papagiannis (Boston: D. Reidel, 1980).

20. R. W. Sinnott, "The Wandering Stars of Allegheny," *Sky and Telescope* 74 (October 1987): 360–363.

21. R. S. Harrington and B. J. Harrington, "Barnard's Star: A Status Report on an Intriguing neighbor," *Mercury* 16 (May-June, 1987): 77–79.

22. S. Beckwith and A. Sargent, "HL Tauri: A Site for Planet Formation," *Mercury* 16 (1987): 178–181. Also, S. Beckwith, "On the Number of Galactic Planetary Systems," in *The Search for Extraterrestrial Life: Recent Developments*, I. A. U. Symposium No. 112, ed. M. D. Papagiannis (Boston: D. Reidel, 1985).

23. D. W. McCarthy, Jr., R. Probst, and F. J. Low, "Infrared Detection of a Close Companion to Van Biesbroeck 8," *Astrophysical Journal Letters* 290 (1985): L9–L13.

24. Note on B. Zuckerman and E. Becklin article in *Nature* (12 November 1987) in *Sky and Telescope* 75 (January 1988).

25. A. Fruchter's "discovery" is discussed in *Sky and Telescope* 75 (May 1988): 468.

26. R. A. Schorn, "VB8B's Vanishing Act," *Sky and Telescope* 74 (August 1987): 139.

27. B. Campbell's claimed discovery is discussed in *Sky and Telescope* 74 (August, 1987): 125.

28. Peter van de Kamp, *Principles of Astrometry* (San Francisco: W. H. Freeman and Company, 1967).

APPENDIX 5

1. Bernard Oliver, "The View from the Starship Bridge and Other Observations," *IEEE Spectrum* 1 (January 1964): 86–92.

2. A. P. French, *Physics: A New Introductory Course*, part 3, *Relativity* (Cambridge, MA: The Science Teaching Center, Massachusetts Institute of Technology, 1966).

APPENDIX 6

1. G. Marx, "The Mechanical Efficiency of Interstellar Vehicles," *Astronautica Acta* 9 (fasc.3, 1963): 131–139.

2. G. Marx, "Interstellar Vehicle Propelled by Terrestrial Laser Beam," *Nature*, 2 July 1966, 22–23.

3. B. M. Oliver, "Efficient Interstellar Rocketry," Paper IAA-87-606, presented at 38th I. A. F. Congress, Brighton, United Kingdom, 10–17 October 1987.

Index

LACOMBE COMPOSITE HIGH SCHOOL LIBRARY